JOYCES MISTAKES

Problems of Intention, Irony, and Interpretation

JOYCES MISTAKES

Problems of Intention, Irony, and Interpretation

Tim Conley

UNIVERSITY OF TORONTO PRESS
Toronto Buffalo London

© University of Toronto Press Incorporated 2003
Toronto Buffalo London
Printed in Canada

ISBN 0-8020-8755-8 (cloth)
ISBN 978-1-4426-1298-3 (paper)

Printed on acid-free paper

National Library of Canada Cataloguing in Publication

Conley, Tim, 1972–
Joyces mistakes : problems of intention, irony, and interpretation /
Tim Conley.

Includes bibliographical references and index.
ISBN 0-8020-8755-8

1. Joyce, James, 1882–1941 – Criticism and interpretation.   2. Joyce,
James, 1882–1941 – Criticism, Textual.   3. Irony in literature.   I. Title.

PR6019.O9Z5273 2003   823'.912   C2002-905607-1

University of Toronto Press acknowledges the financial assistance
to its publishing program of the Canada Council for the Arts and the
Ontario Arts Council.

This book has been published with the help of a grant from the
Humanities and Social Sciences Federation of Canada, using funds provided
by the Social Sciences and Humanities Research Council of Canada.

University of Toronto Press acknowledges the financial support for its
publishing activities of the Government of Canada through the
Book Publishing Industry Development Program (BPIDP).

Contents

Figures vii

Acknowledgments ix

A Note on Texts xi

I PORTALS OF DISCOVERY: AN INTRODUCTION 3

1 Re: Cognizing Error 5
2 The true scholastic stink 14

II WRITING ERRORS 21

3 Fault Lines: Representing Modernism's Errors 23
4 Multiple Joyce Questions 40
5 Fickling Intentions (I) 59
6 (Sic) of irony 81

Intermittences of sullemn fulminance 95

III READING ERRORS 99

7 Performance Anxieties 101
8 Fickling Intentions (II) 118
9 The allriddle of it 134

Erroneous Conclusions 149

Appendix: Quashed Quotatoes 153

Notes 157

Bibliography 175

Index 187

Figures

1 *Dubliners* (Penguin) 48   66
2 *Finnegans Wake* (Penguin-Viking) 214   78
3 *The Norton Anthology of English Literature* (vol. 2) 2311   79

Acknowledgments

Custom recommends that here should appear some statement of ownership for the mistakes that reside within this book. To make such a claim would be entirely antithetical to my purpose: 'my' errors have a greater hold upon me than I on them. Whatever else they may be, the mistakes of the following pages still seem to me to have come about – as Gerty MacDowell would put it – 'accidentally on purpose,' but they could not have been made without help.

The chapter entitled '(Sic) of irony' benefited from an early draft reading and subsequent discussion at the 'Hides and hints and misses in prints' panel at the London Joyce conference in June 2000: I am indebted to Greg Downing, Andrew Mitchell, and Sam Slote for the interest and acumen they brought to that occasion.

Parts of this book have appeared (in slightly different forms) in the journals *Midwest Quarterly*, *Papers on Language and Literature*, and *James Joyce Quarterly*, and I owe their respective editors thanks.

In its guise as a dissertation, this project was supervised by Jed Rasula, whose inspiration abides. The enthusiasm and care shown by my editors at the University of Toronto Press have made the publication of this book a pleasure.

For their various gestures of assistance, support and friendship, I wish to thank Stephen Cain, Brian and Brenda Conley, Michael Groden, Paul Joyce, Bob Perelman, Allen Ruch, Sylvia Söderlind, Asha Varadharajan, and especially Clelia Scala.

A Note on Texts

Abbreviations

CW	*Critical Writings*
D	*Dubliners* (Penguin edition)
D-C	*Dubliners* (Cambridge Literature series, ed. Andrew Goodwyn)
FW	*Finnegans Wake* (Penguin's 1976 issue)
GJ	*Giacomo Joyce*
L I	*Letters I*
L II	*Letters II*
L III	*Letters III*
P	*A Portrait of the Artist as a Young Man*
SH	*Stephen Hero*
U	*Ulysses* (Penguin edition, follows the 1960 Bodley Head edition)
U-C	*Ulysses* (*The Corrected Text*, ed. Hans Walter Gabler)
U-F	*Ulysses* (Orchises Press's facsimile of the 1922 first edition)
U-M	*Ulysses* (Modern Library edition)
U-RE	*Ulysses* (Picador's *Reader's Edition*, ed. Danis Rose)
U-V	*Ulysses* (Vintage edition, follows the 1961 resetting)

Although I have adopted for citational conveniences the abbreviations sanctioned by institutions such as the *James Joyce Quarterly*, I have not seriously endeavoured to establish any edition as authoritative for my use (let alone anyone else's). Readers will note, for example, that the Penguin *Ulysses* principally consulted here is based upon the 1960 Bodley Head edition, not the most academically esteemed in the hierarchy of troubled texts. In the course of the following analysis the whole family is reunited in comparative surveys, though this process sometimes results in scuffles between profligates and favoured sons. The rule of my choices here is the spirit of the argument: *faute de mieux*.

References with asterisks (*) are to the appendix.

JOYCES MISTAKES
Problems of Intention, Irony, and Interpretation

I
Portals of discovery: An Introduction

Who can understand his errors?

(Psalms 19:12)

Which of us can control our scribblings?

(James Joyce, in conversation with Arthur Power 89)

CHAPTER ONE

Re: Cognizing Error

> And then he starts with his jawbreakers about phenomenon and science and this phenomenon and the other phenomenon.
>
> (*U* 394)

There is a standard joke about modern art in which an abstract painting is incorrectly hung, usually upside down. As stale as the gag is, the persistence with which it is adapted in films, television, homogeneous *New Yorker* cartoons, and so on connotes more than simply a perceived schism between lowbrow gallery groupies and untalented but high-priced hucksters (depending on the side of the divide from which one elects to view it). Can 'art' be 'wrong'? Can even the quotation marks in this question be removed? Pity the curious philistines, coax those who will not face the question: they know not what they see.

The question needs to be addressed, if not comprehensively answered, and itself needs to be questioned in turn. Insofar as it is possible, this investigation will limit its focus to literature, and more specifically to the integrity of text. 'Misreading,' the ideological bête noire of many deconstructionists and the hobbyhorse of critics such as Stanley Fish and Umberto Eco, needs to be matched with its counterpart possibility, 'miswriting.' The terrain of the investigation elects itself; for the works of James Joyce are, themselves, in so many ways the crooked paintings: a skewed portrait, a smutty novel, an illegible singularity.

Inauthenticity dwells in the details. Nowhere is the struggle between intention (author's or reader's: the latter is sometimes forgotten) and effect more volatile within a text than at the fault line of error, or the perception of error, since neither of the extreme positions above can recognize 'error' as a concept. Joyce's teasing interplay between the 'meaningful' or 'meaningless' nature of words in each of his works, but especially in *Ulysses* and *Finnegans Wake*, suggests that extremes meet: this principle of co-incidence, I argue, ought to be instrumental in considering Joyce's authorial position and process.

The discussion that follows at least purports to have a thesis, and it is this: Joyce's aesthetic 'progress' occurs apace with his appreciation and integration of error as a principle of composition and publication. Yet such a thesis necessarily depends upon a series of specific hermeneutic axioms that, given its central position here as well as its possibly contentious connotations, deserves serious scrutiny. These are the very general problems of my subtitle: problems of limits. Ostensibly a respectable academic argument, the following is more truly an inquiry in the mode of thinkers like Wittgenstein ('investigations') and McLuhan ('probes'). For the moment, however, let us dwell on the second word of the title, 'Mistakes,' and leave the unpunctuated pluralities of 'Joyces' for the second half of this introduction.

In *The Pound Era*, Hugh Kenner asks: 'Is the life of the mind a history of interesting mistakes?' (230). If this seems a salient question as it is posed, it could be made more piercing with the removal of the apparently synonymous frames, 'life of' and 'history of': is the mind made of/by interesting mistakes? Rational perceptions seek frameworks of order, formulae, patterns and, in turn, construe these as the phenomenal norm. If the fascinating, albeit somewhat radical, theories of scientists like Roger Penrose concerning human consciousness's connection with the workings of quantum mechanics possess any tenability at all, our tacit disavowal of entropy's anti-design in and on all things – including those faculties of perception – may be a serious blind spot for many of the studies of cognition. My own critical concern with reading, with literary experience, is for me one of these relevant studies, and I think that the notably textual orientation of the phrase 'margin of error' signals literature's own (oft-neglected!) integral subjection to randomness, besides articulating the fact of error's exile to the ends of the page.

The sciences attribute a value to the notion of error from which the arts skulk away. However much, and often as rightly as, authors have taken to task C.P. Snow's demarcation of two 'cultures,' certain differences in perspective recur. Mistakes assume a central role in the study and practice, the give and take of mathematics, say, or microbiology, wherein hypotheses can be indifferently advanced and retracted, but the student of the humanities is all too often loath even to recognize the fallible side of theory, wedged as he/she may be between pressures of performative rectitude and the current attractions of a blinkered relativism, one of the more inauspicious elements of 'postmodern' thought, at least in its more vulgar incarnations.[1] (These are generalizations, to be sure, but tenable ones. Rarely does a theorist in the humanities acknowledge either intellectual detachment from or the probably low mortality rate of his or her theory, despite the example of Stephen Dedalus's unconvinced *Hamlet* 'algebra.') Italo Calvino specifically characterizes the difference between the 'cultures' – a word which by now seems to have all but spent its usefulness – as linguistic: 'Scientific writing tends toward a purely formal and mathematical language based on an abstract logic indifferent to its context. Literary writing tends to construct a system of values in which every word, every sign, is a value for the sole reason that it has been chosen and fixed on the page. There could never be any meeting between the two languages, but (on account of their extreme disparity) there can be a challenge, a kind of wager between them' (37).[2] This is one wager that I take up in this book, and there will be more bets to come in the chapters ahead (see, especially, Pascal's in chapter 5), some of them longshots and 'throwaways.'

Although there is discussion of 'form' in all of the arts, and many great works of art suggest, trace, or push against the inherent restrictions of these respective forms, the concept of error has traditionally been a stumbling block for aesthetics. Aristotle recognized error as a deviation from recognized forms, a want of 'technical correctness,' but doubted 'whether the error is in a matter directly or only accidentally connected with the poetic effort' (*Poetics* 2338). Little has changed in this pattern of thought by the time George Steiner, for example, arrives to ponder 'what are the "truth-functions" in music, in what sense can a musical instrument be said to be "true"? (True to what?) ... In music there can be violations of the declared contract with a chosen, rule-bound form such as a fugue or a canon. These can be labelled "errors" in a technical-conventional matrix. The beginner gets his exercises in counterpoint "wrong"' (*Errata* 71). Steiner's inconclusive wrestling highlights an important duality in the expression of 'wrong.' He has in mind here the German sense of *'falsch,'* untrue, and consequently, he tends to conflate rather than compare the 'wronging' or 'wrongful use' of art to art's "wrongness." Unfortunately, the badly hung painting is not (much) touched by this approach. It may be a shame or even a travesty that art, and maybe artists and audience, too, have been so hard done by, but what 'wronging' and how it is manifested are left as occult mysteries.

The attribution of a moral grade to error is not, it is worth noting, a habit of the modern sciences, but singly and sometimes pathologically one of the arts. Its practice as well as the vocabulary all too regularly employed when making assessments of this or that 'canon' signal the theological pedigree of hermeneutics and its sometimes regrettable endurance. (I will have more to say about the modernist experiment as a shift away from this limiting paradigm in chapter 6.) I am not suggesting that such conscientious questioning, *wie ist Kunst falsch*, is not valuable or even not perhaps imperative – and, to be sure, for their part the scientific ethicists might do well to emulate it – but rather I confine myself to remarking on the way in which this form of interrogating 'error' is exclusive and usually represents an evasion of the basic but obviously difficult problems of interpretation.

As things are, the discourse of science offers the clearest and most intriguing history of error as an idea, and one that is actually fundamental to the discourse itself. In his *Novum Organum* (1620), Francis Bacon reflects on the capacity for error in human observation and reckons that 'human understanding is like an uneven mirror that cannot reflect truly the rays from objects, but distorts and corrupts the nature of things by mingling its own nature with it.' In addition to this general capacity for errors, 'each of us has his own private cave or den, which breaks up and falsifies the light of Nature' (54).[3] Bacon is, of course, outlining the limitations and directives of natural science, but his ideas presage various hermeneutic theories forwarded and debated within the academic literary studies of the past half-century, and it is difficult not to associate that 'uneven mirror' with Joyce's 'cracked lookingglass' (*U* 6), more the signifier of the (an) author's modus operandi than a symbol. Although Bacon's rationalist program compels him to denounce error(s), as well as the weakly human propensity for proliferating them, there are intriguing instances when he relaxes in the attacks, or at least demonstrates a curiosity about the connection of intentionality or authority to error. Valuable experience, 'if taken as it comes, is called accident; if it is deliberately sought, it is called experiment' (91), but there is no denial of happy accidents. (Again an irrepressible Joyce association: the ever-mutational 'felix culpa' of *Finnegans Wake*.)

Bacon is rather more ambivalent about distortions of text: 'It will also happen, no doubt, that someone, after reading my natural history and tables of discovery, will find in those very experiments some things which are not quite certain, or downright false, which may make him think that my discoveries depend on foundations and principles that are false and doubtful. But this is not important, such things are bound to happen at first. It is only as if in writing or printing, one or two letters were misplaced, which does not impede the reader much, since the mistakes are easily corrected from the meaning ... No one therefore should be troubled by these mistakes I have described' (120). The observer of natural phenomena may very well err, but error in a text is somehow mitigated or circumvented by 'meaning,' something Bacon assumes is apparent. Whether this faith is placed in literate sensibility (like a suspicion of print's novelty), some enduring luminosity of apodictic truth(s), or a compensatory kind of 'narrative coherence' (Lecercle 22) is not altogether clear. What is certain for Bacon is that texts are less erroneous than is human interpretation.

By the decade following the publication of *Novum Organum*, crafty Galileo Galilei had formulated a series of precepts on observational errors. These precepts contend '(1) that errors are unavoidable, (2) that small errors are more likely than large ones, (3) that measurement errors are symmetrical (equally inclined to overestimation as to underestimation), and (4) that the true value of the observed constant is in the vicinity of the greatest concentration of measurements' (Bennett 90).[4] These are the presentiments of probability, the initial clearing of a path to be followed by Planck and Heisenberg. It is also a discernible shift away from the frequently inhibitive provision of authorship, since the observable nature of phenomena challenges the ineffability and mystery customarily retained as simultaneously the prerogative and proof of a creator.

Einstein's objection to quantum mechanics' provision for the governance of chance – 'God does not play dice' – is arguably the most potently fallacious formulation of the century (perversely within the same pantheon as 'the war to end all wars' and even '*Arbeit macht frei*'). Within the decades of scientific discourse since this pronouncement, there have been many rejoinders (e.g., Niels Bohr rebuked Einstein for telling God what to do), but the comment most germane to the following exploration of creative and interpretive error(s) is Stephen Hawking's 'not only does God play dice, He sometimes throws them where they cannot be seen.'[5] Where modern science's determinism is continuously questioned at the same time as more and more 'grand' theories are developed, the problem of intention confounds the study of literature, despite challenges (inspired by modernist thought and by authors like Joyce) to the foundations of the idea of authorship, such as those of Heidegger, Barthes, Foucault, and so on. The common fallacy of a literary text's status I want to renounce here is one comparable to that of Aristotle's view of the nature of the universe: that it (the universe of the text, the text of the universe) is as it appears, and always has been and always shall be thus. Of special note is that no creator is required for this situation. Neither is there any framework for gauging error, since both error and its recognizer require that non-existent authority for their own expression. From this point of view it is unthinkable that, as *Finnegans Wake* puts it, 'the compositor of the farce of dustiny makes a thunpledrum mistake' (162.02-3): all such terms are emptied of value.

A better conception (i.e., more adaptable to greater considerations and contexts) of the author/text relation might analogically be called more or less Newtonian, in that it might be posited as an 'advancement' beyond the Aristotelian view. The universe/text, though it has an origin, from which certain theories and postulations concerning its formation can be derived by careful observation, is subject to limited, discernible forces of change. Order is the structural focus, however, and 'within or behind or beyond' (*P* 215) this order is understood to be an intelligence, an intention. This creator may very well be paring his (or her) fingernails, because 'after' the act of creation, to borrow an image from Heidegger, 'the artist remains inconsequential as compared with the work, almost like a passageway that destroys itself in the creative process for the work to emerge' (263). Authors may be dead, they may be social constructs, but they do not play dice.

An even more exciting notion of creation and authorship removes the linear finality these concepts have been presumptuously assigned. (Heidegger, in the well-known essay from which the above is taken, has difficulty with the problem of the 'work's' temporal existence.) The text is always being written, creation is ever a 'work in progress.' Within this possible world – the Heisenbergian view? – moments of disorder are not only possible; they are entirely probable.

Analogy goes only so far, of course, and scientific discourse neither can nor should be co-opted for other than instructive points on greater appreciation of its difference, but with the examples of Galileo and Bacon in mind, I want to suggest a provisional outline of kinds of perceived[6] *written* error. This exercise ought to be considered as an imaginative topographical survey of a fabled distant land, as impressionistic and fundamentally rhetorical as the preacher's meditation on hell in the third chapter of *A Portrait of the Artist as a Young Man*, but useful for these qualities as grounds for argument (though I hope for the reader's sake that this exercise is rather less torturous and perversely ridiculous than that sermon). Merely stipulating here their qualities – this is the beautiful lesson of both Frege and Borges – does not by any means confirm that such ideas of 'error' are of any legitimate reality (whatever that is). The types of error catalogued here are constituted by plausible cases and instances in which a discerning reader's head might shake. Such a hypothetical reader is 'discerning' because of his or her awareness of a weighty body of evidence, or of an ability to provide a proof, that effectively contradicts or negates whatever is perceived amiss. What follows, then, is really a list of aberrations to accepted rules, frameworks, and concepts.

First, the most obvious kind of mistake to be noted in a literary discussion. Within the set of *errors of syntax* are grouped the banes of those pundits whom Steven Pinker calls 'language mavens' (370–403): generally, errors of spelling, grammar, and punctuation. There is nothing close to a sufficient lack of examples in the world of print at large to warrant any being manufactured here.

Errors of calculation, by slight contrast, include those instances in which one discovers a statement like $2 + 2 = 5$. This type can be closely associated with syntactic errors, since mathematics as notation has, itself, a distinct grammar dependent upon consistency, although it is neither a regularly clear, precise categorization nor exclusively mathematical. The myopic student who squints at the question $5 + 5 = ?$ may supply an answer of 25. Is this, strictly speaking, an error of calculation? The fast-grading computer, of course, has no doubt, but the human-held red pencil pauses and hovers above in intelligent sympathy. It must be stressed that the gradations between the kinds of error crudely anatomized here blur as soon as intention is considered.

A third language-based error that might be considered concerns anomalies in word selection. Neglecting the most basic agreement in a shared language, by whatever extreme (from calling a dog a 'cat' to calling it a 'borogove'), results in *errors of vocabulary*. Translation, in its most pedestrian usage, also offers openings for this sort of error (calling a dog *'un chien'* or a *'madra'* is, we might say, less 'wrong' than calling it *'ein Ferkel'*).[7]

Where errors of syntax, of calculation, and of vocabulary depend upon the presence of a fairly arbitrary[8] but continuously operational force of correction lurking in the background, there is a much larger but inescapable need for a light of rectitude in which to see the shadows of errors of empirical fact, or *historiographic errors* (Joyce: 'and their eyes are darkened for the errors of men go up before them ever as dark vapours' [P 249]). These errors are essentially deviations from an established collection of knowledge. Keats, for example, notoriously miscasts 'stout Cortez' rather than Balboa as first European witness to the Pacific, and Kafka's Statue of Liberty holds high a sword rather than a torch (*Amerika* 3). Is Keats submitting to the pressures of sonnet form? Is Kafka winking at a nation he never visited? Again, intention is the crucial note of doubt.

There may also be discernible *errors of continuity* in a text, a general phrase borrowed from film studies. Movie buffs are tickled by comparing editing flubs that betray the difference between the represented sequence of events (logical or chronic) and that of the representation's production: discernible daylight in nighttime, dinner plates that replenish themselves, gladiators with wristwatches, and so on. A good example of the textual variety is the infamous incorrect ordering of two chapters in the New York edition of James's *The Ambassadors* (see Butler xix–xxi). Obviously, any argument for recognizing these instances as errors requires an acceptance of a meta-logic to the textual representation of a continuous and at least potentially coherent reality. Surrealist literature, for one, dodges this categorization entirely.

This taxonomy-in-miniature has been a very nice little exercise in the New Critical style. However, it is also done in what the psychologist James Reason not altogether unjustly calls 'a broad but shallow fashion' (ix). Obvious problems abound, even if the phantom of intention can be kept at bay (which, for better or worse, it cannot). Yet, like any working model, this 'imaginary map' of textual error is chiefly useful for its flaws – after all, the best thing about building sandcastles is attacking and decimating them afterwards. It is obvious that my hypothesized schema of types is nowhere near comprehensive and demonstrates the tenuousness of each categorization within it. (If Keats had cast 'stout Balboa' in the role usurped by Cortez, the line would not scan, and this would be registered as a type of error not listed here, a kind of *performative error*, and instance of parapraxis within what Steiner has termed the 'technical-conventional matrix.') Where rigid aesthetics manifest themselves, neither poet nor critic can help but disassemble them. This is a point that I will carry further, to propose that this urge may be acted upon within one's own performance (in this case, Joyce's and ultimately mine), in the chapters ahead.

Some fairly facile but significant principles – or at least valuable working analogies – can be evinced from this example exercise in recognizing textual error. The first idea is very basic: error *can* be systematized and grouped, if only in part. This point is very important, since it at least negates the discomfiting possibility that slips, flubs, and deviations are of such a chaotic nature that no logic or language can grasp them even slightly. Wittgenstein's caution should not be forgotten, however: 'there is no sharp distinction between a random mistake and a systematic one. That is, between what you are inclined to call "random" and what "systematic"' (57). Errors can be generic, systematic, even wholly predictable – at least, if we are 'inclined' to consider them so.

In this regard, 'art' and 'error' are immediately comparable. Within an artistic discipline there are recognizable classifications of form (in music, for example, a sonata is distinct from a concerto) which are, of course, tenable only insofar as a rubric of 'correct' characterizations stays in place (coolly separating music from noise). At the same time, both impulses to artistic creation and error, whatever else they may be, are polymorphous and transgressive. Whether error is creative, or even can be, is not evident from this analogue but remains a possibility vital to the discussion that follows.

The final and perhaps the most self-evident of these principles is the governing philosophy of this study's method; as Robert Duncan would have it, 'let us / begin where I must // from the failure of systems' (847). Theory itself is a study of errors – typically those of other theories, if not (ideally) its own. Literary theorists, however, too often aggrandizing concepts rather than questioning them, are usually the last to admit this. Characteristically stern on this point, Steiner rejects what he sees as the misuse of the term 'theory' in the humanities and discerns 'a profound logic of sequent energy in the arts, but not an additive progress in the sense of the sciences. No errors are corrected or theorems disproved' (*In Bluebeard's Castle* 135). The concession to a 'logic' in the arts allows for refutation, since any logic is inherently a corrective force, and profound logic is the nursery of profound deviations. With the suggestion that all critical 'systems' are not simply suspect but wonderful to attack and disassemble, I turn now to the errant author who inspires these acts, James Joyce.

CHAPTER TWO

The true scholastic stink

For all of his Aquinan schemas, Stephen Dedalus has a problem:

> –*If a man hacking at a block of wood,* Stephen continued, *make there an image of a cow, is that image a work of art? If not, why not?*
> –That's a lovely one, said Lynch, laughing again. That has the true scholastic stink.
>
> (*P* 214)

By June 1904, though, he seems decided upon the point:

> –The world believes that Shakespeare made a mistake, [Eglinton] said, and got out of it as quickly and as best he could.
> –Bosh! Stephen said rudely. A man of genius makes no mistakes. His errors are volitional and are the portals of discovery.
>
> (*U* 243)

If any portals are clearly discernible at this moment in *Ulysses*, they open onto a good number of questions from Joyce's reader. What exactly are these 'portals of discovery'? Where does one find them, and for whom are they open?[1]

A relevant truism bears repeating here, because too many commentators overlook it when confronted with these scenes. The judgments of Stephen are by no means either those of Joyce nor a final '*point bien visible*' (Joyce's letter to printer [Groden et al., *James Joyce Archive* 21:140]), an end to debate. What Stephen's problem does emphatically reflect is Joyce's fascination with the issue. As for Stephen's positional switch, this is hardly the only instance where he appears to retune his considerations of a problem in the fancy-inspiring hiatus between *A Portrait* and *Ulysses*. Haines is told that he beholds in Stephen 'a horrible example of free thought' [*U* 23], a quip that many critics have transcribed with Haines-like appreciation, but the speaker in younger days also announced that 'there is no such thing as free thinking inasmuch as all thinking must be bound by its own laws' (*P* 187). Joyce's consciousness is as litigious as it is lexical, tracing along, as it does, the concentric boundaries of rules within rules. In his Hegelian way, Joyce tangles himself within nets before he flies past them.

In the discussion that follows I examine a range of Joyce's writings to establish the emergence of an awareness, an aesthetic of error, but ultimately all of the streams of discourse here will flow into the *Wake* ('vund vulsyvolsy' [*FW* 378.30–1]). There is a clear sequence of literary self-awareness in Joyce's publishing history, wherein each new text emerges as a meta-text reassessing those that have appeared prior to it. It seems that Joyce has anticipated (a cautious choice of verb for so early a point in a discussion of authority and chance) the intertextual game of snakes and ladders criticism plays within his oeuvre. This phenomenon has been remarked upon often enough, but what about the matter of textual integrity (revisions, variations, errata, pirated editions)? I will return to this question.

The overwhelming body of criticism that exalts the significance of *Ulysses* causes many to lose sight of the novel's insignificance and of its own exaltation of that quality. Itself both a winning throwaway and a feast of crumbs, *Ulysses* poses as the 'useless' text, just as *Finnegans Wake* later represents itself as an 'epical forged cheque' (*FW* 181.16). Joyce, ever fond of renegotiating Wilde's most artful epigrams, styles his supreme fiction as supremely useless.[2] This effort is also – perhaps perversely – tributary to *The Odyssey*, since Roberto Calasso expresses 'the nerve of Homeric theology' as the aphorism '[i]n the maximum pointlessness lies the maximum splendor' (340). For all its encyclopedic vigour and all its author's oft-cited testimonies to its irrefutable veracities, *Ulysses* is a record both of erroneous details and of the record's own incompleteness. (I will refocus on this second aspect in Part II.) The '[s]tately, plump' book, like the phenomenon of consciousness that it masterfully emulates, ends almost in spite of itself. Molly's 'yes' is as much a resignation as an acceptance; it is the note of authorial systematization's end. In fact, it is at last the answer to the young Stephen Dedalus who asks: 'But was there anything round the universe to show where it stopped before the nothing place began?' (*P* 16). As much as it is an archive, *Ulysses* is emphatically a fiction too, and these separate impulses guarantee that all of the data presented to the reader are in a state of negotiation and flux. The exercise of immodality, so readily eluctable to the closed mind of one such as the Citizen, is apparent not in Stephen but in Bloom. Patrick McGee observes in Bloom 'a kind of dialogic space, an open zone like a magnetic field into which every word that comes along tends to insert and double itself in presenting itself as "useless," in resisting the relation to things and standing out in relation to the other words that its differential position excludes. These relations become visible in Bloom's discourse because Bloom refuses to exclude anything or make a decision about the final value of a word; he produces a surplus of linguistic value because he never subordinates one discourse to another in a hierarchy of representational forms' (*Paperspace* 79–80). Bloom's absorbent consciousness is the novel's own best analogue for its composition (though I will later make a case for Molly's 'deranging' editorial presence). While this may be no startling suggestion, it is a point of foundation for many arguments that appear in the following chapters. Joyce's private admission to Beckett of his intuition that he may have 'oversystematized' the book (Ellmann 702) does not mean that it is pristine or faultless. In fact, since every system has its failures and errors, a complex conflation of systems allows for a greater traffic of compositional mistakes. It is this availability to distortion in *Ulysses* that makes it a superlatively relevant case study for this investigation of modern concepts of error.

There is also the beleaguered publishing history of the book to consider. Joyce's haunting promise to keep the professors busy meant more than a scrambling pursuit for workable meaning(s); ultimately it entails a paper chase not unlike any of those run by ad-struck Bloom. Editing, legitimating, and 'correcting' *Ulysses* often seem to be more contentious enterprises than reading, analysing, and discussing it. Indeed, many scholarly careers have been built on this principle: John Kidd has made a career out of arguing with Hans Walter Gabler, for example, and a competing series of projects clambers to assemble respectable electronic texts of the Joyce canon. Some of these initiatives – particularly the new 'scandal' of Danis Rose's 'Reader's Edition' of *Ulysses* – will be closely regarded in chapter 5. While it does initially seem odd to reflect that what is so often cited as the century's greatest novel is its most notoriously unstable and contested (literary) text, beginning with the next chapter I will argue that the two qualities are not only reconcilable but effectively co-dependent and co-supportive.

Of course, typography runs wildest in Joyce's last book. The author of *Finnegans Wake* was able to smile back upon 'his usylessly unreadable Blue Book of Eccles' (*FW* 179.26), a telephone directory never to be installed in any booth, a comprehensive guide to that which never was, and a uniquely nerdy traveller's guide good for one day of the year. Its 'sequel' is more process than product, novel less in noun than in adjective.

The study of *Finnegans Wake* can be likened to a game of pinball: the reader fires into the book, trying for as many resonating ricochets, valuable points of contact, and rescues from oblivion (game over!) as possible; and – relief for the desperate – in the end, 'begin again' (625.32), there's always another ball. Consider the high scores obtained by the mnemonic rebounds from the relevant example of the word 'err.' Counting homonyms alone, one finds 'ere' (occurring 48 times in the text, not counting its possible incorporations into portmanteaux), 'air' (33), 'hair' (46), 'e'er' (9), 'Eire' (4), 'heir' (2), and so forth. ('Err' itself appears 6 times.)[3]

This analogy, as well as its weaknesses, demonstrates the logographical difference between the *Wake* and Joyce's earlier works. Where exact maps could be imposed upon the wanderings of *Ulysses*, for example, and structural analyses of Dante served to frame the tale of Tom Kernan's salvation, the ball is always rolling 'by a commodius vicus of recirculation' (*FW* 3.02) through the unglossabilities of the interminable *Wake*. Eloise Knowlton is correct to chide Roland McHugh for his double standard in *The 'Finnegans Wake' Experience*, rejecting the guidebooks and blazing his own trail into the text before turning 'to add to the corpus of "substitutions"' (Knowlton 3). Reading the *Wake*, Knowlton comments, 'is a project of colonization, of taming a wilderness, of turning a 'jungle' into a 'map.' Of marking off borders. That the text of [*Finnegans Wake*] resists this kind of separation (in short, is neither plant nor animal) does not ground a reevaluation of the project of taming, but only a greater sense of challenge in doing so, a deeper joy in discovering so lush a subject for critical cartography' (5).[4] There is tongue in Knowlton's cheek. Her book, *Joyce, Joyceans, and the Rhetoric of Citation*, is an expression of uneasiness with the Joyceans' 'delicious submission to Joyce's style' (111), if not with the very problem of 'quoting' some form of privileged source text and/or author in the first place. The idea and principle of the *résistance du texte* are germane to my purposes here, but the somewhat diffident accusation of her conclusion[5] – that the autonomy the 'quoting' critic of Joyce affects is bogus – has a hopelessness to which, I think, few could really subscribe: 'When critical practice demonstrates a transgressive melding of method or style, this breach of quotational separation is intentionalized or ignored, and a declaration of independence made. For Joyceans, the denial itself demonstrates a continuity it attempts to reject. Caught in its own readerliness, unable wholly to fulfill the writerly duties of quotational criticism, the uneasy modern critic lives out an ambiguity latent within his attempts to fend off its object: he struggles to keep it at a safe distance, even as it sustains him as food' (112). Leaving aside the bizarre construction of this last sentence (to what exactly does 'its' refer?) and the gendering of 'the uneasy modern critic,' this suggestion of 'independence' is specious, and it is this fact that the 'transgressive melding' of Joyceans not only points out but celebrates. Yes, strangely, it is less difficult to speak of Joyce and his works than to write about them. However, the quest of 'quotational criticism' (the textually engaged) is quixotic, if this adjective may be glossed as a dialectic between the impossibly foolish and the inestimably virtuous. For a scholar, there are far worse situations.

Although I have elsewhere stated my own dissatisfaction with the major body of *Finnegans Wake* criticism as it stands, a brief summary bears repeating here, with some further reflections on the problem of representing the book that, perhaps like no other before, actively moves against representation.[6] The challenge Joyce's last book poses to criticism's tendency towards allegory, as well as to any attempt 'to be strictly literal' (FW 575.12–13), has not been directly accepted. Rather, the response has been evasive, a surrendering recourse to paraphrase, sometimes to gloss half-heartedly, to cut 'skeleton keys' and produce bombastic titles like *The* Wake *Lock Picked.* (*Ulysses* warns, 'Love laughs at locksmiths' [U 474], and the *Wake* playfully echoes: 'Shshshsh! So long as the lucksmith. Laughs!' [FW 148.32].) It is not that attendants of the *Wake* (and who could claim to be anything more?) have been too 'abcedminded' (FW 18.17), but that they have not been nearly 'abcedminded' enough. The unchecked urge to simplify, to reduce what may be more than metaphor to something less than metonymy is ridiculed by Joyce's language's own self-awareness, most vibrant in the *Wake*. Where 'definitions' are offered, their face value is momentary, since their appearance is immediately afterwards (in the printed sequence of characters) very probably due for alteration, mutation, any kind of orthographical defacement.[7] Where this does not happen, the repetition of a word is like Gertrude Stein's lexical laundering efforts at a fast spin cycle: 'Talis is a word often abused by many passims (I am working out a quantum theory about it for it is really most tantumising state of affairs). A pessim may frequent you to say: Have you been seeing much of Talis and Talis those times?' (*FW* 149.34–150.01). To ask 'what is Talis?' one might as well ask 'what is this?' ('This': this word, this mottled page, this question, this existence.) Joyce's 'quantum theory' leaves the critics still in the Newtonian mindset sitting under an appletree – instead of contemplating 'the only abfalltree in auld the land' (*FW* 88.02).

Error, contends Roy Gottfried, 'is not possible in *Finnegans Wake*, where all meanings are potential' (21). If this is so – and it is far from apparent – why the persistence of Joyce and his revolving retinue of amanuenses with proofs? (I will have more to say about this last word in Part II.) Why does the book meditate with such lemniscate design about the erroneous nature of the human mind and soul? Joyce makes his 'thunderous mistake' (*FW* 509.09) loud if not clear, and the proposition of an all-meaning text co-exists, with explicit endorsement from the *Wake*, with that of a 'meaningless' one. Between the possibilities of all readings being valid to none, there are also in flux the possibilities of misreadings. This flux, both paradox and dialectic, is what makes the *Wake* such an unavoidable challenge to literary interpretation and, as Terry Eagleton notes, a trial by fire for any hermeneutic theory one cares to advance.[8]

Finnegans Wake, I argue, is not only stammering but (proto-) stochastic. Oliver St-John Gogarty's 1939 characterization of Joyce's loosing of the word from its sacramental confines – '[t]his arch-mocker in his rage would extract the Logos, the Divine word or Reason from its tabernacle, and turn it muttering and maudlin into the street' (4) – marks a critical road little travelled. Gogarty's maudlin Logos importantly resounds in Margot Norris's writings on the decentred narrative of the *Wake*,[9] and Jed Rasula has rightly enhanced the image of the stravaiging signifier to the letter as character ('*Finnegans Wake* and the Character of the Letter' 517–30; the misdirectioning of letters begins in *Ulysses*: think of how the 'l' in Martha's 'world' seems to have leapt there from Bloom's name in the newspaper). While the book madly integrates meaning(s) – or at least seems or purports to do so – it disintegrates itself.

Michael Groden ('Contemporary Textual and Literary Theory' 259–86) has argued convincingly for the need to confront or overcome the wide divide between the respective practices of (abstract) literary theory and (applied) textual criticism. More specifically, Thomas Jackson Rice despairs that those 'who have been "boring into [the] mountain" of *Finnegans Wake* from the top down, analyzing its grand themes and meaning, have yet to meet those who have tunneled into the novel from the bottom up' (113). This book is not intended as thematic criticism, though, to be sure, in the pages that follow there will be a complementary pattern of text-drawn thematic associations with mistakes for the purpose of demonstrating the problem's prominence and polymorphousness for Joyce. Obviously, the tropes of sin and fallenness have no little importance in discussion of this subject, but as my focus is basically hermeneutical, I have elected here to adopt only at propitious junctures relevant metaphors for direction.[10]

Another blind spot: the usual psychoanalytic apparatus for dealing with instances of parapraxis will not be taken up here, at least not for the purpose of delineating character or the psychological novel. Neither is this study to be a compendium of irregularities, but of course a significant number and variety of examples will be concentrated upon (see the appendix).[11] Instead, this project – an inevitably self-reflexive one, like any textual analysis of textuality – represents an attempt to answer, both by argument and by example, Groden's call for a constructive convergence of methods.

Derrida's wise remark about the naïveté of claiming to 'have read Joyce' ('Two Words for Joyce' 148) serves as a hinge for the swinging door between the central two parts of this study (II, 'Writing Errors,' and III, 'Reading Errors'); just as we are always reading Joyce and have never finally 'read' Joyce, so is the process of creating *Ulysses* and *Finnegans Wake* not a fait accompli – any more than the propagation of error itself has an end. We, as readers, must try to learn from Joyce's mistakes, so that we might make braver ones.

II
Writing Errors

Text: open thy mouth and put thy foot in it.

(*U* 752)

Our wholemole millwheeling vicociclometer, a tetrodomational gazebocroticon (the 'Mamma Lujah' known to every schoolboy scandaller, be he Matty, Marky, Lukey or John-a-Donk), autokinatonetically preprovided with a clappercoupling smeltingworks exprogressive process, (for the farmer, his son and their homely codes, known as eggburst, eggblend, eggburial and hatch-as-hatch can) receives through a portal vein the dialytically separated elements of precedent decomposition for the verypetpurpose of subsequent recombination so that the heroticisms, catastrophes and eccentricities transmitted by the ancient legacy of the past, type by tope, letter from litter, word at ward, with sendence of sundance ... as highly charged with electrons as hophazard can effective it, may be there for you

(*FW* 614.27–615.08)

CHAPTER THREE

Fault Lines: Representing Modernism's Errors

In his thoughtful and cogent *Preface to Modernism*, Art Berman posits the aesthetic movement as a wide-ranging critique of modernity's failure to live up to its utopian, 'progressive' promises, particularly those engineered by empiricism. Appositely, I argue, the most rigorous and stylish of these forms of critique pursue failure as an application and sometimes, as in the notable cases of Broch's *Der Tod des Vergil* and Joyce's *Finnegans Wake*, as subject for protracted, highly style-conscious meditation. As the *telos* of being seems to recede faster and faster – in tandem, in fact, with the retreat of mythic moments of origin – so the ends of narrative are stretched, frayed, and made impossible or redundant. Even the most rudimentary story endings are negated. 'Happily ever after,' the promised bliss of consummation, is steadily demolished by the sagas of James and Mann. The eponymously assured deaths of Broch's Virgil, Beckett's Malone, and Joyce's Finnegan(s) are deaths only of expression (the rattle at the end of the printed alphabet), as the language of the book expires.[1] As a technique or mode, the fragmented, unfinished work is in part a renewal of ambivalent experimentation with textuality by the Scriblerians in the eighteenth century (by this time the affiliations between Sterne and Joyce or between Swift and Wyndham Lewis are more or less critical commonplaces, but it is as easy to think of an institutionalized Smart compounding his *Jubilate Agno* as of a caged Pound pressing on with more *Cantos*). This qualified return is fuelled by a dissatisfaction with the nineteenth century's impositions of standards upon, and subsequent, comfortable absorption of, textual narrative, but the new century had other compelling anxieties to and with which writers responded. The bravest of these writers questioned not only their own individual authority but that of the act of creation itself.

Pre-modernist frameworks were being challenged not only by modernist artists themselves – it is ridiculous to ascribe all the contagious modes of innovation of any era to its artists alone – but by revolutions within empirical disciplines. It was during the middle years of the *Wake*'s composition that Edwin Hubble discovered that the universe was expanding[2] and Kurt Gödel outdid Bertrand Russell's paradox by producing his startling incompleteness theorem, a proof against proofs.[3] For Joyce both discoveries would have uninterruptedly synthesized with his reading of Giordano Bruno, who posited an infinite space in which could be contemplated the necessary convergence of opposites ('God es El?' the *Wake* wonders, without providing a direct answer [246.06]). But whereas the particularities (and peculiarities) of Joyce's approach to these contingent new realities may be unique, the assimilation itself is not. If 'modernism is a movement that concerns itself with defining value, not with facts,' as Berman argues, then '[t]he scientist can dominate the method, if only the artist and the writer dominate the context' (25). The Futurists, the Surrealists, the Dada assembly all could adopt and adapt (albeit selectively) growing, exciting mythologies for their respective aesthetics. Accelerating galaxies? See Marinetti. Entropy as direction? Regard Tzara.

Invoking Adorno's claim that 'totality is the lie' and the examples of unfinished Proust, Musil, and then focally Benjamin, George Steiner briefly outlines the radical shifts in social tectonics that disallow the foundations for niceties of artistic completion: 'The accelerando and violence of recent history, the large-scale disappearance of the privileges of privacy, of silence, of leisure that underwrote the classic practice of reading and aesthetic response, the economics of the ephemeral, of the disposable and recyclable which fuel the mass-consumption market, be it in the media or in the factory, militate against enactments of completion and totality' ('Work in Progress' 3). Creative sufferers of the 'modernist crisis' (Steiner, *After Babel* 189) pit the product of text (self-contained, authoritative volumes for predictably bourgeois consumption) against the process of writing (flawed, continuous, sometimes even allegedly 'unreadable'). After poets such as Mallarmé, the process of poetry appears as 'one of calculated failure: characteristically, a modern poem is an active contemplation of the impossibilities or near-impossibilities of adequate "coming into being"' (Steiner, *After Babel* 190).

This invocation of 'being' (incidentally the verbal pivot of Irish-English syntax) rightly points to the fact that there are also reasons which might be loosely grouped as philosophical for the modernist-fostered belief that sublimity resides in the contemplation of 'the endlessnessnessness ...' (*U* 355; an important instance of infrequent ellipsis in Joyce's later writings). Within avant-garde modernism, the tendency away from completion that I am delineating here bespeaks an awesome desire for a new pact between author and reader, a shared experience of the sublime. 'A novice painter asked his teacher, "When should I consider my painting finished?" And the teacher answered, "When you can look at it in amazement and say to yourself '*I'm* the one who did *that*!"' Sartre curtly glosses this parable in his next paragraph: 'Which amounts to saying "never"' (Frechtman 27: 'Un peintre apprenti demandait à son maître: "Quand dois-je considérer que mon tableau est fini?" Et le maître répondit: "Quand tu pourras le regarder avec surprise, en te disant: 'C'est *moi* qui ai fait *ça*!'" ... Autant dire: jamais' [Sartre 51]).[4]

I offer a few important examples of such 'novices' and their idiosyncratic disavowals before ultimately bringing focus to *Ulysses* and *Finnegans Wake*. Although the authors and texts chosen for this purpose are done so as to suggest a non-exclusive genealogy – a paradoxical 'progression' of sorts, if you will – of modernism's interaction with error, the 'fault lines' I am charting here provide an analogical complement to the 'phases' of modernism outlined by Berman. Contending that modernism 'is more an aggregative self-reflection than an order' (30) and that chronological grids cannot be evenly imposed upon the fluctuating 'phases' of aesthetic progress, Berman locates 'first, the maintenance of the *transcendental ideal*, the continued reliance on the numinous substratum of ordinary reality that the urban world itself eagerly relinquishes in favor of city life. There is, second, the maintenance of an *ethical ideal*, a personal commitment, exercised through an aesthetic, in a society whose shared ethical bonds are rapidly being replaced by ... structures that function not as a universal ethics ... And there is, third, the *aesthetic ideal*, the insistence on the autonomy of art, as a self-sustaining realm fully satisfying the artist, the world at large now being incapable of so doing ... Late modernism, a fourth phase, summarizes, competes with, reduplicates, parodies, and also exhausts the earlier phases' (30; italics in the original). I am borrowing liberally from Berman's thought to show how each such 'ideal' in turn affects the modernist's rendering of a subject and how this act of textual representation confronts its own inevitable distortions and limits, given these ideals.

Melville, the only novelist writing in English besides Joyce (at least before the Second World War) to grapple seriously with the problems posed by Sir Thomas Browne's prescient formulation of the encyclopedic project's ready collapsibility,[5] gives his Ishmael some very modernist reasoning: 'small erections may be finished by their first architects; grand ones, true ones, ever leave the copestone to posterity. God keep me from ever completing anything. This whole book is a draught – nay, but the draught of a draught' (Melville 157). The ineffability of the written subject is given full force in *Moby-Dick*, as the whale is demonstrably and terrifically superlative. Physically and mythologically oversized, mysteriously submerged and freakishly unique (as white as an untouchable blank page), Melville's leviathan ultimately ravages its pursuant would-be circumscribers and leaves chastened the necessary witness 'floating on the margin' (625). Melville's narrative is in no small degree an argument against itself. Although there are 'those disheartening instances where truth requires full as much bolstering as error' (223) – a remarkable phrase – the refutations of 'Monstrous' and 'Erroneous' pictures of whales swell to include the author's own: 'any way you may look at it, you must needs conclude that the great Leviathan is that one creature in the world which must remain unpainted to the last. True, one portrait may hit the mark much nearer than another, but none can hit it with any very considerable degree of exactness. So there is no earthly way of finding out precisely what the whale really looks like' (289). (It is easy to see how, in this respect, a single Thursday in Dublin is very like a whale.)[6] Berman observes in 'early modernism' at least 'the possibility of outright desperation' (31), and there is more than just possibility in Melville's novel: the opening counsel to 'Give it up, Sub-Subs!' (Melville xxxix) has an aura of hopelessness, an impossible surrender that anticipates Beckett's 'I can't go on, I'll go on' (*The Unnamable* 414).

Melville would have thanked you not to call him a transcendentalist; the term had for him a disagreeably Emersonian taint. With Hawthorne, he would disagree with the rejection of evil as a palpable force in favour of a more benign abstraction of spiritual unity. And yet, as explicitly as *Moby-Dick* does not offer an ideal of elevating consciousness or spirit, the ultimate simultaneous smash of the ship, quest, and novel is entirely revelatory: 'man's insanity is heaven's sense' (454). The 'realism' of mimetic, written narrative does not by any means capture the 'real' of the Leviathan, but the struggle is the narrative discourse.[7] Roland Barthes contends that 'as soon as the writer ceased to be a witness to the universal, to become the incarnation of a tragic awareness (around 1850), his first gesture was to choose the commitment of his form, either by adopting or rejecting the writing of his past' (*Writing Degree Zero* 4). The case of *Moby-Dick*, however (timely in 1851), is one of a noteworthy difference. The 'tragic awareness' of both Ishmael and Melville stems from a recognition of (1) the impossibility of serving as true witness to the universal, and (2) the impossibility of not serving as a witness at all. Any rejection of 'the writing of his past' is only prologue to the rejection of the author's present writing. (Again, the anticipation of Beckett, but also of Kafka, is here.)

From whaleness to snailness. It is a degree of difference that propels Marianne Moore from Melville's (doomed) vision of totality to her own diminishing vision of the slightest flash of consciousness, of beauty, of 'the genuine' ('Poetry' 36). 'If "compression is the first grace of style,"' ('To a Snail' 85), she has it; if 'it is human nature to stand in the middle of a thing' ('A Grave' 49), human nature fails in the comparison with a snail. Contrary to the wide-angle lens of her American predecessors Melville and Whitman,[8] her affections are for the minute and momentary. 'Where Proust and Joyce add and add,' summarizes Andrew J. Kappel, 'Moore cuts and cuts' (126).[9] Indeed, the problem of not writing, felt so acutely by and in Beckett and Kafka, is presaged by Moore's reluctant lyricism. Moore's infamous statement about her *Complete Poems* – 'Omissions are not accidents' (vii) – is as determinedly austere and pointed as any of T.S. Eliot's or Gertrude Stein's statements. Moore's compulsion for largely subtractive revision is seemingly implacable, and her poems need to be considered as variations rather than as intractable edifices (or Shakespeare's 'gilded monuments'). The most notorious example of her method's ferocity is the textual history of 'Poetry.' The poem initially expands from thirteen lines (as it appears in *Observations* [1925]) to thirty-eight lines (as it appears in *Selected Poems* [1935] and in so many anthologies that elect this form of the poem), but then shrinks to a lonely three (as it appears in *Complete Poems* [1967] and – this only begins to give the lie to the title – in subsequent editions of this collection).

Because of these revisions, *Complete Poems* is a haunted text, replete with what Kappel terms 'expressive voids' (135). Just as Emily Dickinson's dashes so often seem to bespeak an absence, to register the unsaid, Moore's truncated verses in this volume galvanize the reader's memory: a spiked dose of déjà vu. Someone has erred! The dismayed reader who consults Clive Driver's 'Note on the Text' to *Complete Poems* finds that it raises more questions than it answers or settles. Custom dictates that he offer the assurance that '[t]he text conforms as closely as is now possible to the author's final intentions,' though we may harbour doubts about the exact implications of 'as is now possible' and even wonder what the adjective 'final' connotes (the mind, Moore insists, is an enchanting and enchanted thing for its fluctuations, its not being 'a Herod's oath that cannot change' ['The Mind Is an Enchanting Thing' 134–5]). Driver goes on to write: 'Five of the poems written after the first printing of this volume have been included. Late authorized corrections, and earlier corrections authorized but not made, have been incorporated. Punctuation, hyphens, and line arrangements silently changed by editor, proofreader, or typesetter have been restored. Misleading editorial amplifications of the notes have been removed' (Moore, *Complete Poems* viii). This, surely, is the curtest outline of editorial apparatus possible, however much it tries to appear as an expansion on the poet's 'Omissions' principle. Yet even Driver's design is not the end of the editorial framework, for the poem's notes, though they may be cleansed of 'editorial amplifications,' are introduced with Moore's own 'A Note on the Notes.' There, she refers to her 'hybrid method of composition' in the present tense – *as a living thing* – and allows that the reader 'annoyed by provisos, detainments, and postscripts could be persuaded to take probity on faith and disregard the notes' (262). In other words, the reader of the *Complete Poems* is offered a sort of editorial agency: more cuts and cuts! The collection's title is disingenuous. In 'Efforts of Affection' Moore articulates her sense of

> wholeness –
>
> wholesomeness? say efforts of affection –
> attain integration too tough for infraction. (147)

These are lines worth contemplating. Here, the narrative revises itself; here, 'wholeness' is never a given, as even the poem's hesitant title suggests from the outset. The attainment – watch how pronounced, how regular is the present tense in her poems – is an ongoing effort. Robin G. Schulze offers Moore's interest in Darwin as a useful source of metaphor for understanding why the poet 'did not "perfect" her poems in a teleological quest for an abstract ideal. She adapted them in response to her changing social, cultural, and textual circumstances' (299). Although Schulze's nuanced argument is a good one, she disdains the word 'progress' for what 'teleological' associations she detects in it (there is no sense of an ending in its etymology – I will say more of this term in specific connection with Joyce in the next chapter) and pays little attention to the impetus for survival as it effects the kind of 'evolution' she describes. 'Authors create texts in order to "do" something,' she concedes, but her assertion of 'authorial selection' as an editorial principle (299–300) still allows that these authors know what the 'something' is, and even how to achieve it. For Moore, the artistic creation, not the artist, 'does'; it is the evolving form and consciousness:

> It comes to this: of whatever sort it is,
> it must be 'lit with piercing glances into the life of things';
> it must acknowledge the spiritual forces which have made it.
> ('When I Buy Pictures' 48)

Or, too, those that are yet making it. The fascination with evolution, natural and poetic, serves as Moore's ethical ideal. Where Melville cannot capture his aesthetic's subject with his method, Moore's method tremulously approaches a near negation of her subject. The white whale and the sparrow-camel, glimpsed, have already vacated the premise(s) of the writing (and perhaps, like Don Quixote, have stepped directly into the literary unconscious) before the reader is done. In fact, it seems strange that neither *Moby-Dick* nor Moore's *Complete Poems* simply disintegrates in one's hands as one turns the final page.[10] Why this is so – why, in a sense, the texts of Melville and Moore fail to fail with distinction – has to do, I think, with their inability to confront the possibility of error (in Moore's term, 'accidents') in anything better than a defensive fashion. Again I ought to point out that these authors are examples, not encompassing generations and peers but rather uniquely exemplifying the incrementally bolder approaches to error and its creative potential. The cases of Melville and Moore are initiatory and, correspondingly, instructive; modernism's greatest mistakes take their lead.

Enter Pound, who boldly aphorizes: 'Motive does not concern us, but error does' (*A B C of Reading* 91). These are the words of a younger poet than the one who opines in Canto CXVI:

> But the beauty is not the madness
> Tho' my errors and wrecks lie about me.
> And I am not a demigod,
> I cannot make it cohere.
>
> ...
>
> Many errors,
>   a little rightness,
> to excuse his hell
>   and my paradiso.
> And as to why they go wrong,
>   thinking of rightness
> And as to who will copy this palimpsest?
>   al poco giorno
>   ed al gran cerchio d'ombra
> But to affirm the gold thread in the pattern   (815–17)

Reading the words, 'my errors and wrecks lie about me,' it is hard not to think of Ishmael amid the debris of the Pequod. In Pound – especially in his later works – there is an implied, active dialectic between this mysterious 'little rightness' and 'writeness,' textuality. Pound actually confronts error as an aesthetic possibility, with different approaches for different kinds of 'wrongness' and with mixed results. The copiers of his palimpsest – authors of transmissional errors – include Pound himself (Froula 139). Two sorts of textual error run through the poem: roughly, mistranslation and historiographic slips. This separation is not observed within the poem, it is important to note, but for the moment let us entertain it for initial consideration.

Much has been said, not a lot of it favourable, about Pound's idiosyncratic translation techniques. While I am not disposed to reckoning here with either the substantial library of extant translation theory or the specifics of modernist contributions and challenges to translation theory and practice, it is important to explore how Pound's 'errors and wrecks' correspond to his decisions as a translator. Is there even such a thing as a 'correct' translation, given that equivalencies between languages are essentially convenient fictions? Kenner offers this answer: 'Time and again the only meaning of "correct" is "traditional." We can sometimes say that a word cannot possibly mean what a translator has written in response to it; more often we can say that he has not written what readers of the original usually understand' (*The Pound Era* 216).

Thus, the reorientation of Chinese ideograms ('[n]o one is going to be content with a transliteration of Chinese names' [prologue to Cantos LII–LXXI 254]) is a form of making it new (itself, Pound argues, an ancient dictum). When a reader encounters the poet's words on the page, he or she finds not the containment of meaning (or Kantian Absolutes), but

>   what is said there
> is rather a character
>   than a true

  ching

  ming

definition. It is a just observation. (Canto LXVI 382)

Pound has one ear cocked always to tradition – though his sinological ventures signify possible conflicts within the intersections of different cultural ideas of tradition – but the other is strictly attuned to his own voice. This splitting of the attention is the commitment to the past confronted with Rimbaud's imperative of and for the modern, and the inevitable discordance of this method is the radical stuff of modernism.

Steiner characterizes the practice of translation (he rejects the concept of a theory) as one of creative mis-taking: 'Poor translation follows on negative "mistaking": erroneous choice or mechanical, fortuitous circumstance have directed the translator to an original in which he is not at home ... Positive "mistaking" on the contrary generates and is generated by the feeling of at-homeness in the other language, in the community of consciousness' (*After Babel* 399). This 'feeling,' it should go without saying, is a construct, but that fact is no invalidation of the effort. Pound's sense of alienation from the America of 'Rosenfeld' (his slur for Roosevelt) and much of English writing compels him to locate himself elsewhere. Juxtaposition expands in his work as both technique and trope. He seems to have actually conceived of idioms as detachable units from cultural lineages, hoping as he did that Confucianism could be positively imposed upon industrial America.

The question of identity's limitations, or the politics of 'at-homeness,' leads to the problem of Pound's (mis-)writing of history. Steiner calls

Pound's 'facility ... to enter into alien guise' his 'divine accident' (*After Babel* 378). Christine Froula has addressed this issue with sensitive intelligence, and I deferentially cite her here at length:

> Pound's demarcation of the historical limits of his own authority within [*The Cantos*] goes a long way toward explaining his tolerance of error in his text – how, for example, he could regard a mistake as a saving sign of the author's ignorance and insist on preserving it. His stance towards the errors in the text reflects a radical transformation of our three-thousand-year-old Western tradition of epic authority, and it is this that accounts for the fact that we still read the poem as though it claims, or ought to claim, the *kind* of authority on which our literary tradition is founded. The historical idealism that leads readers to value correct facts over grasp of historical process, to imagine anyone – the author or ourselves – as capable of taking up an objective vantage point outside history, is also what leads readers to expect from Pound a superior authority that the poem, in representing itself as a poem including history which is itself included in history, is at pains to reject. (162; italics in original)

Pound's embrace of history, the textual reproduction, as flawed an enterprise of any of the multitude contained by the idea of its subject, matches an attitude expressed by Octavio Paz: 'Western history can be seen as the history of an error, a going astray, in both senses of the word: in losing our way in the world we have become estranged from ourselves. We have to begin again' (87). Yet it would be wrong not to know the past, that wealth of enriching faults. Conjunction is the beginning of Pound's *Cantos* – from 'And' to 'that' in the indicative first Canto – leading to steady parataxes; so much so that whether there are more nouns than conjunctions in the whole poem is a very good question. (It is here, moreso than in his short early poem 'A Pact,' that Pound's peace with Whitman is truly and fully manifested.) In his examination of modernity's iniquities, Pound employs 'no "Aquinas-map," no idea or philosophy from which to trace a governing design' (Froula 153) – though unfortunately, his bigotry sometimes assumes such a role – and thus he tries to loosen his hand from hegemonic constraints as it redraws history.

Included in and perhaps central to these constraints are proprietary concerns. Pound challenges the legitimacy of claims to ownership, and it is this challenge that unites his effects of history and translation, since (a) language is the necessary basis of (a) history. To claim to own history, to have a hermetically sealed, critically guarded conception is for Pound unconscionable and ultimately untenable, owing to the threat of poetry. (Canto LXVII cheekily begins, 'Whereof memory of man runneth not to the contrary' [387], an allowance from a poet badly and too often painted as simply a rabid dogmatist.) The same holds true for language, which, when reduced to an object – a complete and refined commodity – is a dead thing. With this understanding, just as Pound can adopt American yokelisms alongside Homeric intonations with the same vigour, Italian and Chinese are borrowed clothes to be tailored, as extensively as the fit may require.

Pound's epic enterprise (as well as those comparable others: Williams's *Paterson*, Zukofsky's *"A"*) may be seen to embody – repeating the earlier quotation from Berman – 'the *aesthetic ideal*, the insistence on the autonomy of art, as a self-sustaining realm fully satisfying the artist, the world at large now being incapable of so doing' (30). The 'errors and wrecks' of the *Cantos*, I suggest, signify this autonomy; for where the collective and institution foster rectitude, the artist commits sabotage. The self-consciousness of modernism, however, leaves its works vulnerable to their own mischief: the 'realm' of art is 'self-sustaining' only insofar as it is self-depleting, too. Eschewing or correcting a mistake is itself a mistake ('faute de ... something more solid / but not in all cases' [Canto LXXIV 463]), so Pound's poem, like a dog trained to attack dogs, reacts against itself. It is even problematic whether the work will tolerate its being called 'Pound's poem'; for as Froula notes, errors in the *Cantos* 'challenge some deeply rooted assumption about correctness' and 'about the ability even to posit, let alone to "recover," an "authorial intention" or an "ideal text"' (7). Jerome McGann similarly suggests that the poem calls the theory and practice of editing's bluff, since 'no final distinctions can be drawn ... between substantives and accidentals, between "the text" and its ornaments – between the work of the poet, on one hand, and the work of the compositor, the printer, even the bibliographer on the other' (*Textual Condition* 147). Hopes for a perfect text, an artefact free of distortion, are oppressive vanities. Even more conspicuously than 'Pound's poem,' the phrase 'free of distortion' is a misnomer; distortion *is* freedom, and it may be not only the right of the poet to err, but also, particularly for the modernist, a responsibility.

'It seems history is to blame' (*U* 24). There is perhaps more irony in Haines's heinous statement in *Ulysses* than has been hitherto acknowledged. Besides revealing the imperialist's absence of mind (how sly of

Joyce to give this unappealing Britisher such an unpleasant French name: *la haine*, hatred), the claim can also be read as a definition: *to make history is to blame*. Blame – unlike 'Bloom,' love's advocate – is etymologically kin to *blasphemy*, the utterance of impieties and profanities, and cousin by association to *blemish*, the error of defect. Joyce shares the epistemological concerns of Melville and Pound, revisions of subject, and he integrates their techniques with the compulsion for auto-revision exemplified by Moore. In the developmental course of his writing history (I mean that in both senses of the phrase) he forges an active aesthetic of error. Subject, language, and typography clash most ferociously in *Finnegans Wake*: 'the pardonable confusion for which some blame the cudgel and more blame the soot but unthanks to which the pees with their caps awry are quite as often as not taken for kews with their tails in their or are quite as often as not taken for pews with their tails in their mouths, thence your pristopher polombos, hence our Kat Krestyberians; the curt witty wotty dashes never quite just right at the trim trite truth letter' (*FW* 119.33–120.04). Please don't mind the 'pees' and 'kews'; they always go astray.[11] Columbus did not find what he claimed, and because one wrong turn deserves another, the 'truth letter' can be rejected and/or relettered as 'trite.'

Retribution, or the redress of grievances, is the mandate and the paradox of modernism's boldest experiments. While Robert Frost calls poetry 'an extravagance about *grief*,' where 'griefs are irremediable' (449; italics in original), Milan Kundera observes that the creator of fiction, especially the novelist, serves as a revenger of history (15–18). From at least Dante (for the poet) and Cervantes (for the novelist) onward, the quest is one of creative revision, wherein the knight-writer, conspicuously errant, tilts against the institution of history as objective and accurate narrative. For the modernist, who can critically look back upon the way in which the rebellion itself has become an institution, exemplified by the pre-packaged format of Victorian potboilers and melodramas, the struggle becomes all the more furious for its new need to be fought on two fronts: against the inevitable mistakes and failures of history as much as against those of literary representation. From this perspective, modernism's convulsions are expressions of its awareness of its antithetical basis; put another way, the body of a modernist text is essentially allergic to itself.

Molly Bloom effectively sums up the recording capabilities of *Ulysses* when she says, 'well small blame to me if I am a harumscarum I know I am a bit' (*U* 927; this is the twelfth and final instance of 'blame' in the novel).¹² Nobody's perfect, in other wor(l)ds – this last syntactic variable being the best example of the principle (Martha's letter), next to the unweavings of 'Penelope.'¹³ Joyce's writings offer an ongoing exchange of blame and responsibility (components of 'awethorrorty' [*FW* 516.19]) between himself as author and his uncertain text. Thomas Jackson Rice suggests that '[w]hat [Joyce] implies in *A Portrait of the Artist as a Young Man* and makes explicit in both *Ulysses* and *Finnegans Wake* is that there is one way out of the dilemma of subjectivity: individuals can escape the subjective, paradoxically, by being objective about their own subjectivity' (Rice 80).¹⁴ Writing is writing under duress, writing hopelessly against itself. The 'constraint' of books like *Ulysses* and *Finnegans Wake* (not discussed as regularly as notions of device, theme, and structure) is a paradox, a privileging of *continuousness* (the motion of the present) over *continuity* (the schema of causality). Like Melville's compendium, Joyce's is a 'harumscarum,' and his method is at least as much a 'hybrid method of composition' as Moore's. In Joyce we find an aesthetic of error that develops more radically than the experiments with narrative accidents in Melville, Moore, and Pound – one that effectively forms a concentrated interplay between those authors' strategies. Stephen's well-known *cri de coeur* in *A Portrait*, besides exemplifying a young artist's fondness for infinitives, prefigures his equation of error with 'portals of discovery': 'To live, to err, to fall, to triumph, to recreate life out of life! A wild angel had appeared to him, the angel of mortal youth and beauty, an envoy from the fair courts of life, to throw open before him in an instant of ecstasy the gates of all the ways of error and glory. On and on and on and on!' (*P* 172). The 'ending' of *A Portrait* is really nothing more than a beginning – for Stephen, certainly, but also for Joyce and perhaps most of all for his readers, whom so much 'error and glory' still await.¹⁵ The further adventures, that walking tour of Dublin and '[t]hat letter selfpenned to one's other, that neverperfect everplanned' (*FW* 489.33–4), represent the apocalypse of modernism; Joyce's trial by error (to borrow Berman's phrase again) 'summarizes, competes with, reduplicates, parodies, and also exhausts' all proximate modernisms. *Finnegans Wake* is an indigestible digest. Fritz Senn notes that 'the verb "err" is ... built right into the first word of *Finnegans Wake*, "riverrun," where indeed it belongs' (*Joyce's Dislocutions* 59; in Philippe Lavergne's French translation of the *Wake*, it is 'erre-revie' [19]).

Instead of an *oeuvre* we find an *ouvroir*, no monolithic form or canon of 'literature' but a realm of 'litterish fragments' (*FW* 66.25–6), in place of a body of work, a 'work in progress.' In conversation with Arthur

Power, Joyce stressed the need for 'an endlessly changing surface' in modern literature, whose creators must 'be prepared to founder' and 'write dangerously' (qtd in Power 95); the lapidary must surrender to liquidity. In his refusal of 'the most pervasive idea of the century in which he was born, the idea of continuity' (Kenner, *Joyce's Voices* 49), Joyce more outrageously than Melville, Moore, and Pound problematizes the stability of text. As a readily domitable physical object, the book is unsatisfactorily limited, 'bound.'[16] When Stephen Dedalus surveys the arithmetical errors of young Cyril Sargent in 'Nestor,' he touches 'the edges of the book' and reflexively thinks: 'Futility' (*U* 33). What Hermann Broch calls 'the necessary completion in the uncompleted' (Untermeyer 98; 'die notwendige Vollendung im Unvollendeten' [Broch 107]) itself is the physicality of the text, that which is in this respect aptly called binding. The useless quality of text *qua* text, outlined earlier in Part I, regularly asserts itself, and again and again in *Ulysses*, within the cycle of its various exchanges of texts (bought, sold, examined, handed out, delivered, posted), the readers within and without the novel face '[c]rooked botched print' (*U* 302). It is very hard to believe, as Roy Gottfried does, that 'Joyce isn't really interested in the process of printing (the few details in "Aeolus" notwithstanding) so much as he is in the product, for the product is what is to be seen' (64).[17] What Gottfried calls a 'few details,' Hayman refers to as 'numerous typographical eccentricities' (*Mechanics of Meaning* 96), while I see them as vital epiphanies of authorial anxiety. 'Almost human' and '[d]oing its level best to speak' (*U* 154), the printing press of 'Aeolus' is sentient and vaguely, comically ominous: 'The machine clanked in threefour time. Thump, thump, thump. Now if he got paralysed there and no one knew how to stop them they'd clank on and on the same, print it over and over and up and back. Monkeydoodle the whole thing. Want a cool head' (*U* 151).[18] The 'cool head' of the writer, the detached modernist pose, is a considered reaction to the technologically reinforced admonishment (or is it a threat?) that 'of making many books there is no end' (Ecclesiastes 12:12). Yet this fancy of Bloom's has a mischievous glee in it unusual for the man always assembling utopian schemes in miniature: the implementation of municipal funeral trams, emergency telephones in coffins for the prematurely buried, and so on. The details of this fancy are subtly resonant. Conspicuous is the choice of 'paralysed,' that 'name of some maleficent and sinful being' that haunts *Dubliners* (*D* 1). 'Paralysis' is the silencing stroke, the vanquishing of the tongue and the mute victory of the immune press. Editor Myles Crawford's fateful words to the telephone in 'Circe' are not 'Stop the presses!' but 'Paralyse Europe' (*U* 585). Father Flynn's lessons to the narrator of 'The Sisters' concern the wonders of the printed word ('and I was not surprised when he told me that the fathers of the Church had written books as thick as the *Post Office Directory* and as closely printed as the law notices in the newspaper' [*D* 5]) and are accompanied by a discomfiting smile in which he 'let his tongue lie upon his lower lip' (*D* 5). The narrator's dark dream features a moistened but stopped mouth on a grey face: 'It murmured; and I understood that it desired to confess something. I felt my soul receding into some pleasant and vicious region; and there again I found it waiting for me. It began to confess to me in a murmuring voice and I wondered why it smiled continually and why the lips were so moist with spittle. But then I remembered it had died of paralysis' (*D* 3). It is fascinating to note how this picture of failed attempts at articulation corresponds with Bloom's listening to the press, and even how the first sentence in this passage is as good a description as any of *Finnegans Wake*, the guilty and stammering text, 'aposterioprismically apatstrophied and paralogically periparolysed' (*FW* 612.119–20).

Joyce's other favourite Irish epidemic, hoof and mouth disease, is also linked with text. Garrett Deasy's wordy letter on the subject appears at a glance to be another conspicuous example of 'bad writing' (*U* 80) in Joyce's novels (though unlike Milly Bloom, whose excuse is her 'hurry,' Deasy has obviously laboured long at the expression of his John Bull–informed prejudices). Incarnations of cows, from 'moocow' to 'Mookse,' signal points of entry into narrative, and Joyce borrows freely on their mythological value. Bloom's locked away cryptogram is described as 'boustrophedontic' in the Random House / Bodley Head edition (*U* 849), a word full of bovine error, and the grazing patterns of the doomed 'Oxen of the Sun' are the historical directions of written prose. When late in *Ulysses*, 'bullockbefriending' (*U* 44 and elsewhere) Stephen instructs or defines 'Text,' as in the epigraph above – 'open thy mouth and put thy foot in it' – his perspective on the 'bull' of text is doubly informed: on the one horn, as a poet who does not seem to write anything himself, he may be seen to avoid (as Plato advised philosophers to do) committing himself to the intransigent and potentially regrettable page; on the other, he has the benefit of his Greek namesake's precedent misfortunes with 'bullfolly' (*FW* 157.07). 'Bull' has for Stephen (and a perhaps less anxious Joyce, too) negative connotations, those of the two masters who seek to regulate his language.[19] John Bull is invoked when Stephen, despairing over the probable inevitability of violence with the soldiers in 'Circe,' mutters, 'Green rag to a bull' (*U* 690), and the potential for excommunication under the aegis of a 'papal bull' is not unlike the poet's banishment from Plato's republic.

Joyce brings the Mad Hatter's unanswered riddle, 'How is a raven like a writing desk?' (Carroll 68),[20] up to modern speed: how are cattle like printing presses? They both have to be tended, prevented from straying. They graze boustrophedontically and each may low in its fashion. Joyce was ever aware that the written alphabet itself is an adapted set of signifiers for beasts of burden (and, where demanded, of sacrifice). Aleph, the protean first letter[21] and for the Kabbalist the point including all points, is Phoenician for 'ox' (Knox 7). Joyce's creative reading of Homer leads him to connect the catastrophes propagated by Ulysses's disobedient (and, it must be said, rather stupid) sailors with the inevitable errancies of language. When it comes to the strictures of language, the slaughter of sacred cows is Joyce's business. His 'most unenglish' (*FW* 160.22) work, Vicki Mahaffey observes, 'reflects a deep interest in the dynamic processes of polarization and reunion that allows systems such as society and language to change, and a serious concern with the pressure to stabilize – and paralyze – such change' (*Reauthorizing Joyce* 4). Paralysis, the foot jammed into one's mouth, is the enemy of the Joycean artist and an effect of the crippling fear of imperfection. Stephen's discontent as a schoolteacher, a marker of student errors, is not unconnected to his inaction as a poet. 'You put your hoof in it now,' Gogarty chides him very early on in the novel (*U* 19). Correction is only a makeshift remedy against the resilient and ever-mutating virus that is language. (All of these medical metaphors will only concatenate in the chapters that follow.) Joyce has diagnosed 'Proof fever' (*U* 154), a false health that bespeaks conformity, stability, and immutability. Accordingly he immunizes his own text against such a state by poisoning it with a modicum of 'all those red raddled obeli cayennepeppercast over the text, calling unnecessary attention to errors, omissions, repetitions and misalignments (probably local or personal) variant *maggers* for the more generally accepted *majesty* which is but a trifle and yet may quietly amuse' (*FW* 120.14–18). Joyce's rarely glimpsed smile may be seen in those last words. It is not at all unfair to say that the *Wake* is *Ulysses*, or all of Joyce's writings, or all literature everywhere, grossly misspelled, mistranslated, thoroughly botched: 'Monkeydoodle the whole thing,' indeed. Although Joyce would criticize Flaubert for flubbing his grammar in *Trois Contes* – 'Il commence avec une faute' (qtd in Ellmann 492) – he himself would open his novel by juxtaposing two near-antonyms ('Stately, plump') and carry on to his 'dud letter' (*FW* 129.07).[22]

Mapping the faultlines is to chart trajectories of modernist experimentation. Text's fallen state – 'the usual crop of nonsensical howlers of misprints' (*U* 752) – is a given that the avant-garde must address and may progressively exploit. Notions of literary texts' stability being the privilege of twentieth-century works are inattentive to these modernists' underpinning forms of rejecting the alleged privilege and, like Moore's 'Pedantic Literalist,' deny a text's 'once spontaneous core in its / immutable production' (37). Observe how even as customarily astute a reader as Guy Davenport can strangely misplace his faith: 'Joyce's texts are what they are – and God will send us a hero to deal with *Finnegans Wake* ... Someday we will have the text [of *Guide to Kulchur*] as [Pound] wrote it. Someday someone will set *Hugh Selwyn Mauberley* correctly. Someday a scholar will reassemble a book of Pound's which our army destroyed all copies of, as it was being published together with Pound's translation into Italian of the *Ta Hio* of Confucius, 'The Unwobbling Pivot' ... Someday we will have a text of *The Cantos*' (*Every Force* 87). Someday our prince will come: the editor as Messiah. (I will have much more to say about editorial agency in the next two chapters.) Unfortunately, Davenport buys into the very utopian teleologies with which the *Cantos* – or is that *The Cantos*? – struggles (and, in Froula's view, ultimately abandons as untenable, even unpoetic).[23] Joyce's texts are *not* 'what they are' but what they are *becoming*. *Finnegans Wake* is not a minotaur awaiting a Perseus (PhD) but a smashed egg from which results a fragmented universe – the four Viconian parts of which could be characterized as 'eggburst, eggblend, eggburial and hatch-as-hatch can' (*FW* 614.32–3) – and Humpty Dumpty will not be put together again ('When will they reassemble it?' [*FW* 213.17]; 'And the teacher answered, "When you can look at it in amazement and say to yourself *'I'm* the one who did *that!*'"'). 'Art,' runs what might be called an incomplete sentence in the *Wake*, 'an imperfect subjunctive' (*FW* 468.09). 'Art' is a verb more than it is a subject or an object. It is continuous – the imperfect tense strangely cooperates with the present formal of 'Thou art Petrus' – and hopeful.

Modernism is not an era, nor is it a closed system. It is an ongoing experiment, a genetic outgrowth. It is a new and happy mistake not eligible for correction, yet one that instructs by experience. Rimbaud stands corrected: *Il faute être absolument moderne.*

CHAPTER FOUR

Multiple Joyce Questions

> I am quite content to go down to posterity as a scissors and paste man for that seems to me a harsh but not unjust description.
>
> (Joyce, qtd. in Ellmann 626)

'Posterity,' such as it is, has had other ideas. The customary portrait of the artist as a monomaniacal auteur is, unfortunately, the most reproduced, especially when it comes to Joyce. I draw here from the materialist-historicist work of Lawrence Rainey as representative example: 'Joyce controlled every aspect of [*Ulysses*'s] production: his approval was required for decisions about paper, typography, cover design, color, even the choice of printing inks. The book was no longer an industrial product shaped by publisher's [sic] conventions and production considerations; it was a token of the authorial self' ('The cultural economy of Modernism' 58–9).[1] In lesser critics this process of image generation/revision quickly turns to caricature, and the aloof and fingernail-paring creator seems more and more like some leering villain of melodrama hatching his Master Plan.[2] Rainey's endeavour to outline the sociology of publication is good in principle – it is in accord with Jerome McGann's fine statement, 'a textual history is a psychic history only because it is first a social history' (*Critique* 62) – but the viability of any of its general conclusions about the experience of modernist publishing is limited. By virtue of its outstanding sales record, a book like Stephen Hawking's *A Brief History of Time* might be (wrongly) assumed by a future historian to have been in its day widely and thoroughly read. Edward Bishop has posed an interesting counter-argument to Rainey's recognition of elitist distribution, based on conjoining examination of antique editions with qualified speculations about *Ulysses*'s readership (as opposed to its mere subscribers).[3] Few writers are so fiercely linked with qualities of detachment and control, both before and after Joyce, who might more readily qualify (say, Blake, or Thackeray) as engineers of Rainey's conspiracy. Too often, commentary takes Joyce at the word of Stephen Dedalus, silently dresses him in a patchwork suit from Eliot and Leavis, and places him for view at a great height: *ecce auctor*.[4]

This, it is well worth noting, is far from the perspective of Joyce offered by his contemporaries. In his 'Analysis of the Mind of James Joyce' chapter of *Time and Western Man*, Wyndham Lewis detects 'vices of style' that he attributes to Joyce's 'unorganized susceptibility to influences' (75). Lewis writes of Joyce like a doctor of a worrisome patient: 'the craftsman is susceptible and unprotected. There are even slight, though not very grave, symptoms of disorder in his art' (90). Ultimately the diagnosis is of a regionalized Cartesian split. Joyce 'has practised sabotage where his intellect was concerned, in order to leave his craftsman's hand freer for its stylistic exercises' (95), but, Lewis concedes, this condition is understandable, since '[t]he intellect is in one sense the rival of the hand, and is apt to hamper rather than assist it' (95–6).

Joyce could be unexpectedly deferential to these kinds of criticisms, taking them by turns quietly in tow for later application in his writing (i.e., adding just more berries to the jam of the *Wake*) or, among intimates, chuckling at their determination to reveal his 'disorder.' In a postscript to a 1921 letter to Robert McAlmon, Joyce laughs up his sleeve: 'Clive Bell wrote an article, I hear, something about modern literature and me. He says that unfortunately I have such mediocre talents as not to justify detailed criticism. What form of suicide do you think I ought to choose?' (*L I* 176). This sentiment – one of many variations of the hard-done-by artist[5] – is echoed in the longest footnote in the 'Nightletter': 'I was thinking fairly killing times of putting an end to myself and my malody, when I remembered all your pupil-teacher's erringnesses in perfection class. You sh'undn't write you can't if you w'udn't pass for undevelopmented' (*FW* 279.F1). Only a selectively coloured hindsight puts Joyce 'in perfection class'; neither he nor his contemporaries, detractors included, think his work without conspicuous faults. Virginia Woolf's early dismissal (we do her wrong to omit that she later changed her mind) of *Ulysses* as the work of an 'undergraduate' has the righting teacher's tone. Dear me, Mr Joyce has got it all wrong.

That is, if he is himself alone responsible. 'A hypnotic fascination with the isolated author has served to foster an overdetermined concept of authorship,' observes McGann, 'but (reciprocally) an underdetermined concept of literary work' (*Critique* 122). In *Multiple Authorship and the Myth of Solitary Genius*, Jack Stillinger remarks how 'the myth of single authorship is a great convenience for teachers, students, critics, and other readers, as well as for publishers, agents, booksellers, librarians, copyright lawyers – indeed, for everyone connected with the production and reception of books' (187).[6] That is, it is congenial to everything but the idea of writing. In his essay 'Wonder did he wrote it himself,' Thomas Vogler gives a hard squint at the 'archaic model of creativity,' according to which a writer 'runs like a machine with only two speeds: on and off': 'The author, however, will turn out not to be a unified subject; he will be a strange diad consisting of an active, conscious, creative pole and a passive, careless copying and proof-reading pole. Only acts emanating from the artistic modality will count as "intentional." The artist is not always a genius surrounded by mere clerks; he becomes a clerk himself when he sits down to write out his own work' (196).* *La traduction des clercs*. One cannot help but think of Little Chandler, always ready to slip away like Clark Kent and re-emerge as T. Malone Chandler, hero of the Celtic note. This transformation is as fraudulent as the isolation; although economics dictate that a 'room' (to borrow Virginia Woolf's formulation), a performance space for that 'artistic modality,' must be obtained even at the cost of time and energy taken from the performance of writing, the truly engaged imagination, the conception of writing, is continuously active. Its beginning and end are not truly discernible, even for the writer. Stephen's tautology in 'Nestor' – 'Thought is the thought of thought' (*U* 30–1) – is a crude miniature of Gertrude Stein's collective work, though both *The Making of Americans* and *Finnegans Wake* are provocative obsessions with thinking of beginnings, and the beginnings of thinking. Although his renderings of it are far sparer than those of either Joyce or Stein, Blanchot gazes just as deeply into this abyss when he writes, 'Pour écrire, il faut déjà écrire. Dans cette contrariété se situent aussi l'essence de l'écriture, la difficulté de l'expérience et le saut de l'inspiration' (232; 'To write, one has to write already. In this contradiction are situated the essence of writing, the snag in the experience, and inspiration's leap' [Smock 176]).

These aporial considerations bring me to the trivial (or perhaps quadrivial) question, the kind found in board games or attached to cash rewards on television game shows,[7] a question with an allegedly demonstrable, correct answer: who wrote *Ulysses*? James Joyce! – in this chapter, I want to shake up this easy formulation of a single author's responsibility for a text and explore the labile 'plurabilities' of 'the event "Joyce," the name of Joyce, the signed work, the Joyce software today, joyceware' (Derrida, 'Two Words for Joyce' 148). In other words, I want to mis-take this factual question and see what portals of discovery open; I want to 'begin again.'

Part of what Joyce suggests by his 'harsh' self-recognition as 'a scissors and paste man' is the fact of his effectively limited role in the creation of his works. Besides the well-studied, Eliotic blueprint for a collage of canonical literary allusions (a cut of Dante matched to a choice slice of Verlaine), Joyce's piracy of others' words extends as far as the casual plunder of newspapers, radio broadcasts, songs, overheard conversations, sermons, his brother's diaries, the names of real people (Artifoni, Carr, etc.); hence the *Wake*'s wonderful admission, 'stolentelling' (FW 424.35). 'In place of style,' Stephen Heath writes of Joyce, 'we have *plagiarism*' (33; italics in original): 'Joyce himself often insisted forcefully on the breaks between his various works (during the writing of *Finnegans Wake* he would ask pointedly to be told who had written *Ulysses*) and that insistence deserves to be remembered. The texts should not be read as the spiritual biography of a full sourceful subject (the Author) but as a network of paragrammatic interrelations constructed in a play of reassumption and destruction, of pastiche and fragmentation' (34).[8] It should be added that during the production of the *Wake* Joyce disavowed knowledge not only of his previous feats of authorship, but even of the one at hand. Eugene Jolas remembers: '"Really, it is not I who am writing this crazy book [*Finnegans Wake*]," he said in a whimsical way. "It is you and you and you and that girl over there and that man in the corner["]' (166).[9] I accept Heath's above claims in general but would colour them a little differently. It is interesting to observe that that unsettling (and unsettlingly italicized) word '*plagiarism*' harkens back to the Latin, *plaga*, which can be translated as 'snare' or 'trap' or especially 'net' – that which at least Stephen Dedalus seeks to fly past. 'Network,' like 'system' (discussed in Part I) and to a lesser extent 'matrix,'[10] has become a catchword for Joyce critics, but it is a misleading choice. That a net is largely made up of negative space, of holes between knotted ropes, is not something Joyce would have us forget, particularly since the *Wake* itself is less a network than an invitation to trace networks around its own vast emptiness. To Heath's forms of dialectic creation, 'reassumption and destruction' and 'pastiche and fragmentation,' I wish to add 'trial and error,' the way by which any labyrinth is explored.

Even if 'spiritual biography' is cast aside, there are plenty of other kinds demanding attention. A multitude of scholarly writings positions the author in a spectacular array of roles: the Irish Joyce, the Swiss Joyce, the Parisian Joyce, the feminist Joyce, the misogynist Joyce, the socialist Joyce, the postcolonial Joyce, the postmodernist Joyce, the filmic Joyce, the multimedia Joyce, and the list undoubtedly will go on.[11] He is never a sum, only parts. The studies of authorship offered in his own writings confirm the extent of his appreciation of this situation. In a 1911 lecture Joyce subdivides Blake into 'the pathological, the theosophical, and the artistic' facets of his 'personality' (CW 220), and in *Ulysses* he plays fast and loose with the profile of the Bard and the problem of attribution, polluted as every character's speech is with the words of other characters and, especially in 'Scylla and Charybdis,' the words of other writers. (Joyce may effectively presage Borges's claim that *Ulysses* precedes *The Odyssey* by hinting that Wilde is prefatory to Shakespeare.) 'Rutlandbaconsouthamptonshakespeare or another poet of the same name in the comedy of errors wrote *Hamlet*' (U 267) is anything but an evasion, irreverent though the expression is. The multiplicities of 'Shakespeare' – 'or another poet of the same name' – are readily accepted by Joyce, whose garbled gospels in the *Wake* are attributed to another composite, Mamalujo, whose signature appears as that of the anonymous illiterate '×' (L I 213). In a 1903 book review Joyce comments that 'a pseudonym library has its advantages; to acknowledge bad literature by signature is, in a manner, to persevere in evil' (CW 112). This last phrase has an unusual sound, because it is the work of the young and inexperienced writer; 'evil' here means something like 'sin against art,' or perhaps more simply, 'error.' But as a whole, the sentence contains a good share of ambivalence, and that seminal word 'signature' is only in the very earliest stages of its growth and bloom within Joyce's lexicon.

Joyce's own proclivities for pseudonyms are sometimes overlooked. There was never any want for companionable examples – Lewis Carroll and Mark Twain are obvious later influences – though his interest alighted especially on Irish examples around him. George Russell, 'Æ,' receives a well-known nod in *Ulysses* ('A. E. I. O. U.' [U 243]), while *Finnegans Wake* pays tribute to Lady Jane Francesca Wilde's 'Speranza' in 'The chape of Doña Speranza of the Nacion' (FW 297.F1) and the wronged spectre of her son Oscar's 'Sebastian Melmoth' as a 'disincarnated spirit, called Sebastion' (535.36–536.01) and 'foull subustioned mullmud' (228.33). Joyce himself appeared in print as Gordon Brown, Stephanus Daedalus, Stephen Daedalus, and James A. Joyce (though *not* Vladimir Dixon, despite the weedy resilience of that Joycean myth) and had various separated, exclusive social identities: James A. Joyce, Mr Joyce, Jim, Papli, Nonno, and so forth.[12]

The multiplicity of identity one finds within Joyce's narratives – the transformations of Bloom in 'Circe' respond to the encouragement, 'Just you try it on' (*U* 561) – is complementary to that found in Joyce's authorship. The universals of everyman and -woman, every place and every time, are always in strange contrast *and* collusion with the particulars of autobiography, while a similar dialectic occurs between Joyce's role as scrivener and the affective powers of his authorial associates. Joyce wrote to Frank Budgen in 1921 about how far his text seemed to wander from him: 'I have dreadful worries about a typist ... She started, but when she had done 100 pp her father got a seizure in the street (a Circean episode) and now my MS is written out in fairhand by someone who passes it to someone else to be typed' (*L I* 159). These 'worries,' like all other anxieties (his own and any others he perceived around him), began to fuel Joyce's texts as subject and strategy, indivisible. Thus the phenomenon of *Finnegans Wake*, where a host of amanuenses, researchers, proof-readers, printers, and what with various meanings might be called unconscious contributors are in motion; charged particles nebulously revolving around a vague central personality. (I will save specific discussion of editing for the next chapter.) These agents had varying – some quite startling – degrees of effect on the generation of the text. It gives one pause to reflect how ready Joyce seemed to be to pass the reins over (Ellmann's word is 'surrender' [592]) to James Stephens, as early into the writing of 'Work in Progress' as 1927. But it is important to remember that while the *Wake* is a conspicuous example of authorial abdications ('Abedicate yourself!' [*FW* 379.19]) within Joyce, it is by no means alone; the entire aesthetic of error (and here 'James Joyce' is the mistake under consideration) is an awareness predicated upon writing and publishing experience. Let me recall the notorious moment, as recounted by Ellmann, of Joyce's dictation to Samuel Beckett: 'in the middle of one such session there was a knock at the door which Beckett didn't hear. Joyce said, "Come in," and Beckett wrote it down. Afterwards he read back what he had written and Joyce said, "What's that 'Come in'?" "Yes, you said that," said Beckett. Joyce thought for a moment, then said, "Let it stand." He was quite willing to accept coincidence as his collaborator. Beckett was fascinated and thwarted by Joyce's singular method' (649). Ellmann's reader may well wonder about whether the unusual word 'thwarted' is Beckett's own term or (as I feel is more likely) the biographer's paraphrase, but 'singular' is very fitting. However, this scene is not so anomalous as it seems.[13] In the first place, moments of 'deputising for gossipocracy' (FW 476.04) occur in previous Joyce compositions, probably beginning with Stanislaus Joyce's arrangement of the verses of *Chamber Music*. In his memoir, *Being Geniuses Together*, Robert McAlmon boldly claims to have 'altered the mystic arrangement of Molly's thought' when he was preparing the typescript of 'Penelope': 'Years later I asked Joyce if he had noticed ... he said that he had, but agreed with my viewpoint' (131).

Knocks at the door have an important place in Joyce and often depict propitious moments of rather Hegelian power-shifting. Stephen is seeking the rarely glimpsed idea of justice in *A Portrait* when he knocks at the rector's ominous 'door at the far end' after his wrongful pandying, and he hears the response, 'Come in!' (P 56). Authority accepts his knock and subsequently transfers authority to him ('You can say that I excuse you from your lessons for a few days ... it is a mistake and I shall speak to Father Dolan myself. Will that do now?' [P 57]). *Finnegans Wake* says, 'Knock and it shall appall unto you!' (FW 528.20), a promise and a threat concerning 'awethorrorty' (FW 516.19). Opportunity is the knock in Joyce, who, as his life and work continued, grew ever more ready to exclude visitors from the former as he was to invite all comers into the latter.

If Beckett and other secretaries are to be sympathized with, what of the phalanx of proof-readers, each of whom felt perhaps the fullest force of the *Wake*'s schizophrenia? José Saramago examines the phenomenon of the proofreader's own multiplicities of self in his remarkable novel *História do Cerco de Lisboa* (*The History of the Siege of Lisbon*):

> O revisor tem este notável talento de desdobrar-se, desenha um deleatur ou introduz uma vírgula indiscutível, e ao mesmo tempo, aceite-se o neologismo, heteronimiza-se, é capaz de seguir o caminho sugerido por uma imagem, uma comparação, uma metáfora, não raro o simples som duma palavra repetida em voz baixa o leva, por associação, a organizar polifónicos edifícios verbais que tornam o seu pequeno escritório num espaço multiplicado por si mesmo, ainda que seja muito difícil explicar, em vulgar, o que tal coisa quer dizer. (22)

(The proof-reader has this remarkable flair of splitting his personality, he inserts a *deleatur* or introduces a comma where required, and at the same time, if you'll pardon the neologism, heteronomises himself, he is capable of pursuing the path suggested by an image, a simile, or metaphor, often the simple sound of a word repeated in a low voice leads him, by association, to organise polyphonic verbal edifices capable of transforming his tiny study into a space multiplied by itself, though it is difficult to explain in plain language what that means. [Pontiero 14])

'Heteronomises' is not a true neologism but an adaptation of Saramago's fellow Portugese writer Fernando Pessoa's 'heteronyms,' deftly constructed poetic identities with individual and interacting names, histories, personalities, and aesthetics. Joyce would have appreciated this form of multiple authorship, since he separates his narrative voices, however indistinctly or elliptically at times, for dramatic effect. (I will return to narrative authorial devices in a moment.) The psychomachia of the 'author' becomes the splitting of the atom (producing the thunderclap) in *Finnegans Wake*. Joyce is, by admission, 'swift to mate arrthors, stern to checkself' (FW 36.35): one writer for this act of writing, another for the next, looking over the shoulder of the first. Just as Saramago's novel grows out of a mild-mannered proof-reader's abrupt and inexplicable decision to 'wrong' a manuscript (i.e., negate what is correct), the *Wake* is an answer to the various 'wrongs' suffered by *Dubliners* and *Ulysses* in their respective extended transmissions.

But who in Joyce's ring ('The ring man in the rong shop but the rite words by the rote order!' [FW 167.32-3]) can be recognized as the 'foenix culprit' (FW 23.16)? Joyceans have been eager to find guilty parties. French printers and prudish typists (or typists with prudish relations) are favourite scapegoats. On the whole, it seems, Joyce's production assistants have got a lot of bad press. Robert Adams reports that 'the printer of *Little Review* was an ignorant, slapdash fellow with the further professional handicap of moral sensitivity' (274), while Jack Dalton complains that *Finnegans Wake*'s 'typists were often dazzlingly incompetent – what less can be said of anyone typing characters onto the roller of their machine, for instance?' (132). Joyce himself did moan to Grant Richards, 'O one-eyed printer!' (*L I* 133), but in a draft of a later letter to Richards, he presents a more ambivalent position: 'You seem to think I have harped too much upon the printer. But is he not the important person? ... really we all – you, your reader and I – are unimportant figures: the printer, alone, is important. His office, apparently, supersedes those of reader and publisher' (*L I* 177). There is still some grousing here within the sly humour but also, and I think more significantly, a pragmatic recognition of publishing's essentials.

This developing awareness goes hand in hand with the progress of Joyce's aesthetic of error. 'Progress' may sound like a strange word to use, but its meaning – to advance, to move forward – is a logical response to the invitation, 'Come in.' T.S. Eliot reflected that Joyce's work needs to be surveyed and understood as a 'journey' or 'progress' (qtd in Litz 121). Phillip F. Herring writes that '[n]o other term but progress will do to describe that process by which we are continually able to learn new things about writers such as Joyce' (78–9). This sentiment echoes Robert Sage's broad claims in his *Exagmination* piece: 'Ordinarily the graph of a writer's career ascends, with slight irregularities, to a horizontal line representing the culmination of development. That is, after a period of trial and error, he achieves an individual manner of expression and his works thenceforth are variations on a theme becoming successively richer perhaps and more perfect but not differing in their bases one from another' (149). Sage then goes on to say that 'Joyce's development, conversely, has been and continues to be a firm mounting line' (149).

This incarnation of 'progress' is that of a nineteenth-century watchword, a cornerstone of Victorian faith in human nature. The steam engine and the dynamo, signs of technological advancement, in this regard were the tangible counterparts to theories of natural selection and class struggle. Although, as Bernard Benstock points out, Joyce 'was suspicious of the ideal of progress as a goal unto itself, without moral basis and a necessary respect for the development of history as an organic entity' (64), Leopold Bloom carries in him some of this ideological strain as an epic-heroic representative of his era. Among his many interests, those in science and civil law bespeak an ongoing concern with social amelioration. Most of the time, though, Bloom's progressive ideas remain unspoken, except when he is directly under interrogation, and the hostility of 'Cyclops' is inverted by the unmitigated, wordy interest in particulars that frames the 'Ithaca' episode. (The narrator of this later episode can be read as another imaginary construct of Bloom, the Spirit of the Age or a historian of consciousness, an ego-appeasing student of one's own thoughts who can represent these thoughts in finer language than can the subject: 'He reflected that the progressive extension of the field of individual development and experience was regressively accompanied by a restriction of the converse domain of interindividual relations' [*U* 778].) But *Ulysses* does not conclude with such ideals. Molly, for whom the past is in the present, is a revisionist to her husband's role as visionary; quite literally, given her revolutions of syntax and punctuation, she is Bloom's antitype. His attachment to specificities, the concern for proper names and effectiveness, are empiricism's scaffolding. Molly the unweaver is the scattered modern mind, content to rely often on 'something' as a blank signifier for a transitory subject. For her there is no such thing as a non-sequitur. The progress of 'Penelope' is entirely literal, in that it is all about going forward, continuing, living within a moment of the living's expression, the time-space in which mistakes are affirmations.

The empiricism invoked in Sage's use of 'progress' is not cherished in Joyce's use of the term; rather, his progress is perversely anti-empiricist, obliterative, a move against strictures. In a 1919 letter to Harriet Shaw Weaver, Joyce writes: 'The word *scorching* has a peculiar significance for my superstitious mind not so much because of any quality or merit in the writing itself as for the fact that *the progress of the book is in fact like the progress of some sandblast*. As soon as I mention or include any person in it I hear of his or her death or departure or misfortune: and each successive episode, dealing with some province of artistic culture (rhetoric or music or dialectic), leaves behind it a burnt up field' (*L I* 129; italics to 'scorching' are Joyce's; the others are added). Although his method of composition is customarily termed additive, Joyce's 'progress' is – to coin a neologism of my own – palimpsestual: every added word or phrase slurs its predecessor(s).

The progress begins early. Flashing hints of his aesthetics-bent interest in the concept of error appear in the line 'Lest bards in the attempt should err' in the 1904 verse 'The Holy Office' (657) and in the 1912 satire 'Gas from a Burner,' which pillories (Irish) editorship for its hypocritical censoriousness ('Shite and onions! Do you think I'll print / The name of the Wellington Monument' [661]), subservience to commercial interests, and generally philistine stupidity. Error is in these instances merely the pejorative label of such minds, but this is only the beginning of Joyce's lifelong consideration of the idea. His incubating fascination with error is more evident within *Dubliners*, where characters trip over words and regretfully wince at their own utterances. The secretarial characters, Farrington of 'Counterparts' and Polly Mooney of 'The Boarding House,' are special, early instances – 'sobsconcious inklings' (*FW* 377.28) – of specifically typographic troubles, the 'hides and hints and misses in prints' of *Finnegans Wake* (20.11). Distracted by alliteration, Farrington writes '*Bernard Bernard* instead of *Bernard Bodley*' (*D* 86; now *there* is an ironic name for an editorial slip in Joyce). When Mr Alleyne hotly demands the missing letters and then an apology, Farrington loses his capacity to forgive the errors of others (the mnemonic counterparts of the English woman's '*O, pardon!*' [*D* 91] and poor Tom's '*O, pa!*' [*D* 94]), commits again the error of misidentification with his own son, and falls to unproductive rage. Polly, the first of Joyce's fictional women to demonstrate an aversion to grammar, who 'sometimes ... said *I seen* and *If I had've known*' (*D* 61), was briefly but significantly an office typist. 'The Dead' is an anticipation and ultimately a contemplation of failure for Gabriel Conroy, whose every utterance inspires immediate anxiety. Talking to Lily, who 'seldom made a mistake in the orders' (*D* 176), Gabriel blushes 'as if he felt he had made a mistake' (*D* 178). Even before he toasts his hostesses, he feels that his 'whole speech was a mistake from first to last, an utter failure' (*D* 179). The pathos of Maria's loveless situation is marked by the lack of reaction to her misrendering of 'I Dreamt That I Dwelt': 'no one tried to show her her mistake' (*D* 102). As the other characters shield her from anything more than a dim awareness of error, even in parlour games, it is left to the narrative frame to supply the cruel answer to why 'it was wrong that time and so she had to do it over again' (*D* 101): the title of 'Clay' is the absent correction.

As a strategy, problematizing the identity of the author begins to merge with questions of what might be called the identity of the text. In *Ulysses* anonymous writings enjoy a vigorous notoriety and circulation, like Denis Breen's provocative but puzzling postcard and '*Sweets of Sin*, anonymous, author a gentleman of fashion' (*U* 868). Due to its own infamy, *Ulysses* had to be disguised when smuggled to nations like the United States, often dressed up in different dustcovers – including one marking the book as the altogether respectable works of Shakespeare. The social function of a title, as well as the entire framework of a book's presentation, was as much in transition in this case as the authorship. The repeated thrusts at Tennyson (always called a 'gentleman poet,' to be compared with the above 'gentleman of fashion') within *Ulysses* partly signify a struggle for the title, but one that Joyce does not choose to 'win' outright. (Leon Edel's oft-quoted description of the author as a collector of grievances is only one side of the coin; Joyce's narratives are incredible, extensive, if sometimes elliptical, records of debts, many of which are his own. Thus, he is less a plagiarist than an erring debtor.) 'Work in Progress' was the name by which readers of early instalments knew the mysterious follow-up book to *Ulysses*, whose 'other' title Joyce liked to have his friends guess at. In a letter to Harriet Shaw Weaver delineating the sigla, glyphs that underline the typographical role of 'character,' there appears a blank square, □, accompanied by the gloss '[t]his stands for the title but I do not wish to say it yet until the book has written more of itself' (*L I* 213). Note the phrasing: as early as 1924 Joyce knew that *Finnegans Wake* was writing – writes – itself. What place has an author in such a book, simultaneously unnameable and continually generating titles for itself both within itself and from puzzled readers? (This list includes neither the less flattering sobriquets Joyce later had for the book in moments of frustration nor that oft-cited but undiscovered volume, *Finnegan's Wake*.)

The distortions of identity (author and title) extend beyond the cover and title page. From these straightforward examples of the function and dysfunction of titles, we quickly appreciate how Joyce invites his readers to seek out narratological dark matter, to consider the negative spaces (those holes in the net mentioned earlier) as significant variables in the hitherto apparently unbalanced equation. In Part III I will have more to say about the impulse among readers of Joyce to explicate; here, I want to examine certain particulars of these explications and see how, rather than why, Joyce 'authors' his works. My argument here is that the shifting narrative views functioning in all of Joyce's works trouble the reader's ability to ascribe words, phrases, and ideas to a single character or entity.

While some of these shifts in perspective can be explained as transitory flashes of borrowed thoughts – what Hugh Kenner once memorably referred to as the Uncle Charles principle, after instances of character-tracing indirect discourse within *A Portrait* – there are many other structural mysteries that Joyceans have stretched themselves to explain. Who 'writes' the title 'Clay,' or *Ulysses*? How or why does Lenehan's 'Rose of Castille' riddle echo in Bloom's head (i.e., 'his' narrative voice) when he was not present for its telling? What does it mean that Molly can apparently address her creator, 'O Jamesy' (*U* 914)?

For the purposes of this argument the recognition of these conundrums is more important than essaying answers, but I do want to give some attention to one of the interesting forms in which criticism has attempted to reconcile these problems. In 1970 David Hayman introduced the concept of the 'arranger' in *Ulysses*, a tantalizing if vague notion of 'a figure or a presence' who 'exercises an increasing degree of overt control over increasingly challenging materials' (*Mechanics of Mean-*

ing 84). A decade later Patrick McGee expressed his doubts: 'Is it another subject whom we can regard as the originator of and proper guide to the labyrinth of the text, that is, the malicious God of this creation? I suspect we are being led off the track by the personification of what is in fact a principle and a power, a principle of arrangement and a power to arrange that which does not originate from a subject – author or narrator – but rather situates the subject. The Arranger [McGee's capitalization] is the structural will to power exceeding the subject, the configuration of symbolic relations in formation, the writing machine that instrumentalizes the subject, who then instrumentalizes the writing machine in reciprocal symbolic exchange' (*Paperspace* 72). The general point here is a good one, though I have some reservations about the use of words such as 'symbolic' (reductive), 'instrumentalizes' (ugly), and especially 'power' (problematic: can power be spoken of as unagented? And in this sense are not power and language the same?). In an afterword to a later, revised edition of Ulysses: *The Mechanics of Meaning*, Hayman observes that this concept provided good subsequent debate, and he restates his vision: 'The arranger should be seen as something between a persona and a function, somewhere between the narrator and the implied author. One is tempted to speak of "him" as an "it," kin to Samuel Beckett's Unnamable, but we are also tempted to think of a behind-the-scenes persona like the shaper of pantomimes, also called the arranger. Perhaps it would be best to see the arranger as a significant, felt absence in the text, an unstated but inescapable source of control' (122–3). The incongruities of 'speak' and 'think,' then of 'see' and 'felt' ('mooxed metaphors' [*FW* 70.32]) here betray the problem with the 'arranger' (re-)arrangement. Whether this mysterious persona/function is operational or has any parallel forms in works by Joyce other than *Ulysses* is not a question raised by Hayman, though I think it is a good one. The unique, controlling agency of Joyce's titles, for instance, manifests itself before *Ulysses* (and what narrative voice collates all of the episodes under this thematically enriching name? Is this the work of an/the arranger?). It is apparent in *Dubliners*, where each title represents a comment on its story. Titles may offer the amplification of a minor word or phrase for diagnostic purposes ('A Mother,' 'The Dead'), an ironic characterization ('Counterparts,' 'Two Gallants'), a combination of both ('A Painful Case'), or, in the intriguing case of 'Clay,' a supplementary fact not included within the text of the story, which also serves as a characterization.[14] The word 'Clay' is the story's missing corkscrew; the story remains sealed without it.

Sequencing may be another 'arranger' effect. *Dubliners*, it has been repeatedly, exhaustively noted, begins with childhood stories and closes with those of maturity and, finally, the words, 'the dead.' *Chamber Music*, on the other hand, appears to have no such clear thematic shape (apart from the title, which in part highlights the accent upon sound, especially the word 'hear' in the poems). The young poet's thirty-six very unexperimental pieces had neither individual titles nor an order of sequence when he delegated editorial powers over such matters to his brother Stanislaus. Does the 'arranger' in this case (if applicable) become an intersection of unstated authorial governances? What happens if a reader ignores the imperious Roman numerals and reads the verses in reverse or random order; is the 'arranger' thus a weaker or less controlling (or reader-restraining) force in this book than in *Ulysses*? I shall return to this question in Part III; at this point, my concern is expressing the possibilities but also the tangible limitations of Hayman's concept.

In response to Hayman, McGee posits a counterpart 'Deranger': 'the Deranger is the limit or other of arrangement – the margin, the space of differentiation, the always imminent possibility of disorder in the world of signifiers' (*Paperspace* 74). What is remarkable about this strategy is the way it runs counter to the urges to declare authors dead on arrival or at least to exile them from their works and hand them a fingernail file. Instead of either privileging the single-gunman magic-bullet theory of the lone Irish mastermind or, conversely, hermetically clearing the text of authorial intentions and 'forces,' the Hayman-McGee debate invites more personae into the cramped space of creation. (And – more knocking at the door – one may think of the overcrowded cabin scene in the Marx Brothers' film, *A Night at the Opera*, where a maid asks Groucho if he would like a manicure: 'No,' he quips, 'come on in.')

In his 1990 study of *Finnegans Wake*, Hayman returns briefly but with a striking difference to the problem: 'On further consideration of that device, I would now characterize the arranging presence of *Ulysses* as predominantly feminine/nocturnal. Thus, in giving control over to mysterious and unpredictable forces, the narrative imposes a simulacrum of irrationality on its activity and gives itself over to the powers of whimsy, which Joyce seems to have identified elsewhere with the feminine/instinctual' (*The Wake' in Transit* 155–6). 'Him or it' is now 'she,' the 'elsewhere' is not specifically located, and what (if any) role an 'arranger' or any similar 'presence' has outside *Ulysses* (say, in the *Wake*) remains unexplored.

With each exchange between Hayman and McGee (as it were, or so I have reset them here), it becomes less apparent that the two writers really agree as to the effect of the 'arranger'/'Arranger,' let alone how to connote or explain the presence of him, her, or it. Moreover, this sexing of the page begins to swerve into the hoary, old argument about whether Joyce's central females are redemptive vessels or whores of Babylon, shapely outlines of order or bawdy voices of chaos ('the necessary disorder of indeterminacy that Joyce felt that a woman's mind could provide' [Herring 178]). Employers of such separations naturally face great difficulties in reconciling the Penelope who weaves with the one who nightly undoes her patterns (completion deferred again).[15] Joyce does sexualize an implied struggle within the act of writing – 'sternly controlled and easily repersuaded by the uniform matteroffactness of a meandering male fist' (FW 123.09–10) – sometimes, but not always, in the exchanges between characters of opposite sex. The word 'metempsychosis' is orthographically disassembled by Bloom, presumably in some attempt to convey his wife's pronunciation, to 'met him pike hoses' (we never do 'hear' Molly's utterance of this word [U 77]), though it is important to remember that his hold on this word is slippery. The word is rewritten, further distorted, and defamiliarized by Molly, who by night-time has reduced it to 'met something with hoses in it' (U 893).

McGee seizes upon Molly's malapropisms as tokens of her (Deranging) authorial influence. Marianne Moore's important word 'omissions' resurfaces: 'Molly gives us just what the doctor (of classical text and reader) ordered – the same doctor who poses to the text of *Ulysses* precisely the question that the medical doctor poses to Molly: "asking me had I frequent omissions where do those old fellows get all the words they have omissions" [U 916]' (McGee, *Paperspace* 179). It is not the object, the matter of 'omissions,' that is contested, but the word. 'Omissions,' such as they are, can be male as well as female: 'Ill wipe him off me just like a business his omission' (U 930), jokes Molly to herself. By virtue of her name alone, readers can estimate the importance of Molly's consciousness as it shapes *Ulysses*. Joyce explained to Budgen in 1920 that 'Moly is the gift of Hermes, god of public ways and is the invisible influence (prayer, *chance*, agility, power of recuperation) which saves *in case of accident*' (L II 147; italics mine). Molly's narratological influence is not quite invisible and does not account for every 'derangement' in the book.

Despite his interest in the narrative disruptions of *Ulysses* and his recognition of Joyce's 'working as much against as with his materials' (*The 'Wake' in Transit* 104), Hayman is surprisingly intransigent when it comes to similar problems of 'arrangement' or its opposite(s) in *Finnegans Wake*. According to Hayman, the question 'who in hallhagal wrote the durn thing anyhow?' (*FW* 107.36–108.01) is 'precisely the reductive sort of thing one does not seriously ask of the *Wake*' (*The 'Wake' in Transit* 45), but this statement pays no mind to the fact that it is the *Wake* doing the asking on this score (see chapter 9, 'The allriddle of it,' for further discussion of the book's questioning agency). McHugh finds in 'hallhagal' suggestions of 'Hell' and the Armenian verb 'khaghal: to play,' but more important, Joyce offers here a mongrel-mockery of 'Hegel' and 'heil Hitler' as another of his gestures of contempt for authorial constructions grounded in empiricism and/or imperialism. If *Finnegans Wake* is 'about' anything at all – and the weakness of this particular precept is always worth observing – it is largely about the hazards of its own construction. Here the term 'metafiction' becomes otiose. No piece of modern writing presents as many caveats, threats (the only pre-*Wake* comparison is Baudelaire's 'Au Lecteur,' and even it is a pale one), and abstruse 'tips' as the *Wake*. 'Now, patience; and remember patience is the great thing, and above all things else we must avoid anything like being or becoming out of patience' (*FW* 108.08–10) – counsel with tongue in cheek.

In attempting to abandon the representation of consciousness, the better to lend the active quality of consciousness, or some semblance of it, to the book itself, Joyce lets his pen be 'transaccidentated through the slow fires of consciousness into a dividual chaos' (*FW* 186.03–6); the alchemy of error. Again and again the *Wake* refers to its own genesis and resultant expansion as a stream of faults flowing from an indeterminable origin (the first fall, the initial thunder, the Big Bang): 'the vocative lapse from which it begins and the accusative hole in which it ends itself; the aphasia of that heroic agony of recalling a once loved number leading slip by slipper to a general amnesia of misnomering one's own: next those ars, rrr!' (*FW* 122.03–6). Er, mutation of 'err' and 'ur,' is a thundergod on the one hand, and the resurrected explicator of the heavens in Plato's *Republic*, a waked Finnegan (Davenport, *Geography* 286) on the other. The *Wake* assiduously resists locating within itself any point(s) of origin, proffering instead 'hints' and 'tips' concerning the erroneous nature of the text's 'progress': 'Diremood is the name is on the writing chap of the psalter, the juxtajunctor of a dearmate' (*FW* 125.06–8). 'Diremood' and 'dearmate' are slips for *dearmad*, the Irish word for 'mistake' (a misprint is *dearmad cló*). Because *Finnegans Wake* suggests a spiral as its own genetic stemma ('juxtajunctor' McHugh glosses as 'harnesser-together'), and credits and/or blames 'the continually more and less intermisunderstanding minds of the anticollaborators' (*FW* 118.24–6) who in distorted forms may themselves appear in the book, isolating 'something between a persona and a function' is like looking for a damp spot in the ocean. All of the 'characters' or 'voices' in the *Wake*, in whatever manner or guise one elects to name them, are writers as much as they themselves are written, each annotating the text/existence of the other ('I'm very fond of that other of mine' [*FW* 408.25]). 'Mr Himmyshimmy,' for example, another sham version of Shem, gives 'unsolicited testimony on behalf of the absent, as glib as eaveswater to those present (who meanwhile, with increasing lack of interest in his semantics, allowed various subconscious smickers to drivel slowly across their fichers), unconsciously explaining, for inkstands, with a meticulosity bordering on the insane, the various meanings of all the different foreign parts of speech he misused' (*FW* 173.30–6). That the drivelling is 'subconscious' and the explaining done 'unconsciously' may suggest that 'the absent' includes the conscious (and perhaps even the consciousness of the) author – Joyce, by any other name. Joyce's endless revision of himself – an attempt to see his own 'Doublends Jined' (*FW* 20.16) – directs his subject and style to merge and lets language and ultimately the idea of an author be merrily 'misused.'

Whatever the failings and inadequacies of the 'Arranger' and 'Deranger' characterizations, the most salient feature of the debate is its form: precisely, that Hayman and McGee feel compelled to *characterize* at all. Re-examining some of the assumptions his early genetic studies of *Ulysses* were based on, particularly that 'a literary work was characterized, even defined, by unity' (Ferrer and Groden 502), Groden sensibly turns to Bakhtin: 'the author for Bakhtin is not incarnated in all or any of the characters. Nor should the author be configured, New-Critical style, as either Wayne Booth's "implied author," posited between the historical writer and the fictional narrator, or as David Hayman's "arranger," located between the implied author and the narrator. Bakhtin collapses the distinctions between author and implied author and author and arranger, but at the same time he avoids the assumptions of a unified authorial consciousness ... Bakhtin's author-as-organizational-center is a meeting-place, a network hub ... anything but a unified whole' (Ferrer and Groden 507). With these considerations in mind, talking of Joyce as an author becomes as slippery an abstraction as talking of his narrative shifts, and 'intention' – the subject of chapter 5 – is a word that works against itself. In *Joyce, Bakhtin and the Literary Tradition*, M. Keith Booker cautions: 'in a writer like Joyce one has to deal with the paradoxical fact that often his authorial intention is apparently that one should not grant interpretive authority to authorial intention, just as Bakhtin's work also emphasizes the importance of the position from which the reader reads' (219). Readers make authors, but Joyce leaves himself always as a work in progress.

Derrida's decision to consider Joyce as an 'event' in 'Two Words for Joyce' is, I think, an especially beneficial option for renegotiating the answer 'Joyce' as well as the question 'Who wrote *Ulysses*?' One of those words that one easily stumbles into tautology trying to define, 'event' manages to contain a lovely cluster of associations. Joyce's use of the word is intriguing, since it is often matched with another, 'shadow,' to describe an inescapable prescience, the antic reverse of Proust's lingering gaze backward. 'Coming events cast their shadows before' (*U* 210) is Bloom's experiential observation; 'If you want to know what are the events which cast their shadow over the hell of time of *King Lear, Othello, Hamlet, Troilus and Cressida*, look to see when and how the shadow lifts' (*U* 249-50) is Stephen's literary one, and the proof of Bloom's.[16]

Joyce is an event, and his own shadow, or other, too. In an article outlining the ways in which Beckett's significant contributions to Joyce studies effectively construct much of what is thought of as 'Joyce' or 'Joycean,' Kevin Dettmar shares an anecdote of his graduate student days: 'my best friend shared with me his own model for the literature of the Romantic period (indeed, for literature, period). Literature, Bill explained, was fashioned from two eternal, antithetical spirits: Blake and "not-Blake" – sort of Blake and anti-Blake, as I understood him, rather than Blake on one side and no one worth talking about on the other. I do not think I realized at the time just how Blakean Bill's framework is, positing as it does that "contraries" make the (literary) world go 'round' ('The Joyce That Beckett Built' 605). Dettmar says that what he did realize was his own 'half-consciously operating under a similar paradigm' (605). Certainly, one could turn over volumes in one's head and consider: Blake or not-Blake, Joyce or not-Joyce; as signifiers of authorial function, these names carry a lot of weight and produce many ripples when cast into a pool. However, where it is, strangely, not too difficult to reckon (in more cases than not) which writers or works are 'Blake' or 'not-Blake,' probably because of that appreciation of inevitable struggle between contraries within Blake's poetic vision, recognizing writers and works of 'not-Joyce' calibre is an altogether treacherous business. For Joyce's Bruno- and Vico-informed poetics, contraries are momentary divisions within a pattern or cycle of disjuncture and reunion. As his aesthetic of error progresses, Joyce – meaning both the identity of the man and the works attributed to him – is as much 'not-Joyce' as he is 'Joyce.'

Is James Joyce the author of *Ulysses*? 'O happy fault! Me wish it was he! You're wrong there, corribly wrong!' (*FW* 202.34–5).

CHAPTER FIVE

Fickling Intentions (I)

It is utterl imposs. Underline *imposs*. To write today.
(U 360)

It is hard to believe in typescript ...
(Joyce, in a 1925 letter to Harriet Shaw Weaver [L I 225])

Any expression of the notion that 'examination of the three terms *final, authorial,* and *intention* will frequently lead us away from an ahistorical conception of the art work toward one of its historical situation and contingency' (Bornstein 8) only means, of course, that much depends upon the terms of that examination. Intentionalism, the interest in an author's conscious and perhaps articulated purpose or design, is a swirling aporia as old as expression. Or very nearly as old, since it is the reaction, the afterthought, the expression of impression. It is criticism's inevitable struggle with determinism and empiricism, in whatever guises they assume: psychoanalysis is only the most recent. (There is a joke about two psychologists passing each other in the street. The first waves and says 'hello' to the second. Further down the street, the second psychologist rubs his chin and says, 'I wonder what she meant by that.') The enquiry of the last chapter was into Joyce's authorial process; this chapter is concerned with textual process and presentation – and, as I am going to frame it here initially, gambling – but it is not a genetic tracking or a comprehensive history of Joyce editions. How does – or can – the presence of editing cohabit text with the possibility of error?

Intentionalism is a form of Pascal's wager, with the outcomes reversed. While the notion of a life free from divine retribution may be thrilling and even probable, Pascal regarded the gravity of the threat of the opposite proposition too great to be dismissed. As Thomas Pepper observes, '[h]ermeneutics as a theological discipline comes into existence because we are fallen, and F.W. Schlegel, Benjamin, and de Man, in this respect, are the inheritors of Pascal' (Pepper 104). Anyone who approaches a literary work, especially anyone with any aspirations to edit the work in question, is faced with a secular version of this soul-shaking gambit: is the work demonstrably and entirely a manifestation of authorial intention (texts, lapidary lex and logos descended from above, can be sinned against) or a product of other, perhaps less determinable factors?

Joyce, for his part, is already ridiculing the sanctity of authorial intention in his uncompleted novel, *Stephen Hero*: 'The hand of Jesuit authority was laid firmly upon that intellectual heart and if, at times, it bore too heavily thereon what a little cross was that! The young men were sensible that such severity had its reasons. They understood it as an evidence of watchful care and interest, assured that in their future lives this care would continue, this interest be maintained: the exercise of authority might be sometimes (rarely) questionable, its intention, never' (*SH* 173). A subtler, more cunning Joyce approached this passage as editor, and chose not to integrate it into *A Portrait*.[1] The idea it represents would re-emerge in his work, however, in more complex forms that would challenge his own would-be editors.

It is not too remedial to ask: What is an intention? Wittgenstein, who could be included among the more profligate inheritors of Pascal, wondered, 'Was ist der natürliche Ausdruck einer Absicht? – Sieh eine Katze an, wenn sie sich an einen Vogel heranschleicht; oder ein Tier, wenn es entfliehen will' ('What is the natural expression of an intention? – Look at a cat when it stalks a bird; or a beast when it wants to escape' [Wittgenstein 165]). That is, intention appears as an urge of desperate nature. E.D. Hirsch, on the other hand, views intention (in this case, in a linguistic act) as an individual's component of a communal creation of meaning: 'Verbal meaning is, by definition, *that aspect of a speaker's "intention" which, under linguistic conventions, may be shared by others*' ('Objective Interpretation' 33; italics in original). For Hirsch there is no 'viable distinction between the nature of ordinary written speech and the nature of literary written speech' ('Three Dimensions of Hermeneutics' 208), or, as his terminology suggests, between acts of speech and acts of writing, at least insofar as the interpretation of these acts is the same process. Reading Joyce, according to Hirsch, would be the same as hearing Joyce speak. Such technologically insensitive arguments are flattened by Jerome McGann's rudimentary observation in *The Textual Condition* that 'the body of the text is not exclusively linguistic' (13). The point to be stressed here is that, because writing is more than a linguistic act, seeking 'intention' involves a wider, more diversified investigation.

Wimsatt and Beardsley conclude their well-known essay, 'The Intentional Fallacy,' whence the now well-worn term, with an unexpected comparison between 'critical inquiry' (understood, of course, to be 'true and objective') and 'bets' (13). Critical interpretation, in their emphatic opinion (and in that of T.S. Eliot), has nothing to do with probability. It is in some ways amazing that such a viewpoint could be so forcefully advanced in the age of late modernism and development of postmodernism, but then it must be understood that literary criticism, then as now, looked enviously at the apparent impartiality and rigour of the fast-moving, specialized sciences (pre-Heisenberg, at any rate).

The statistical analysis of Virgil's metre in the *Aeneid*, performed by Francis Ysidro Edgeworth in the late nineteenth century (see Bennett 116), probably seemed an irritating, almost philistine gesture to literati outside his new and developing field of mathematical study, but current textual scholarship often operates on similar principles. One runs the occasionally surfacing manuscript of alleged Shakespeare, for example, through a gauntlet of algorithms, counting keywords, constructions, and distinct idiomatic turns that constitute the after-the-fact signature of the bard. Determining 'Shakespeare' – or 'Joyce' in the case of the letters of protest included in *Our Exagmination* – involves a critical intuition applied to probabilities. Naturally, this historicization is garlic to the vampires of New Criticism, but the hostility displayed to the idea more significantly shows adherence to a common misunderstanding about studies in probability. Such studies are all about predictability, and this approach has a greater potential currency, I believe, in 'fallen' textual studies than in contemplations of authorial intention alone. Vicki Mahaffey notices in Joyce a strategic deactivation of 'intention': 'If we look to Joyce's texts for evidence of his intentions, we discover him minimizing the importance of authorial intentions by stressing the ways in which they are modified and refracted by the variable processes of writing, transmission, and reception. Joyce, then, uses his authority to recontextualize that authority against the broader backgrounds of history and production, insisting upon the irreducible oscillation between intention and circumstance' ('Intentional Error' 181–2).

Offering an instance of error as an argument against interpretation's dependence upon authorial intention is a straightforward tack. Jerry R. Hobbs recounts an instance of a printer's slip in a newspaper account of the *Pioneer 10* spacecraft's voyage. 'Toward the end of the article, the writer intended to say, "Pioneer 10 carries a message ... in the form of a plaque designed to show ... the place and time where it began its long journey." Instead the newspaper printed, "Pioneer 10 carries a message ... in the form of a plague designed to show ... the place and time where it began its long journey." The fact that what was printed does not correspond to any author's intention in no way diminishes our enjoyment of it, and it is hard to see how we could enjoy it if we did not first interpret it, that is, determine what it means' (11; ellipses in original). Where the 'long journey' of *Pioneer 10* is a given, the transmissional voyage of text from home planet to possible alien readership can often be overlooked. Method or medium necessarily intersects with the 'intention' of the informational message.

That a cat stalks a bird (to go back to Wittgenstein) out of some primal instinct does not fully explain how the bird is killed. (In 'Calypso,' Bloom thinks about such behavioural mysteries: 'Wonder is it true if you clip [the bristles] they can't mouse after. Why?' [*U* 66].) Putting it another way, the fact that the prey dies does not automatically mean either that the predator killed the prey or that the predator's intention to kill the prey has any direct, causal connection to the death of its prey. To edit, then, is to build the better mousetrap, for want of being able to create a cat. *Finnegans Wake* stares at the stains of its own nocturnal emissions and half-jokingly attempts to decide whether the spills constitute literature: 'Whatever do you mean with bleak? With pale blake I write tintingface. O, you do? ... So you did? From the Cat and Cage. O, I see and see! In the ink of his sweat he will find it yet. ... Weeping shouldst not thou be when man falls but that divine scheming ever adoring be. So you be either man or mouse' (*FW* 563.15–33; ellipses added).

Although Joyce should not be confused with a card-carrying surrealist because every style, traditional and emergent, attracts his incorporative interest, concepts such as 'automatic writing' and other attempts at creatively invoking the unconscious are important to bear in mind when retracing his works. André Breton's convulsive beauty finds its place in Joyce's 'endlessly changing surface' (qtd in Power 95): 'All art in a sense is distorted in that it must exaggerate certain aspects to obtain its effect ... Our object [as artists] is to create a new fusion between the exterior world and our contemporary selves, and also to enlarge our vocabulary of the subconscious as Proust has done' (Joyce, qtd in Power 74). A person is given a pencil and told to write down a 'random' word. It runs contrary to all shades of received Freudian wisdom that this can, strictly speaking, be performed. Even William S. Burroughs, who with Brion Gysin enthusiastically investigated 'cut-ups' as a post-Dada technique of narration, pointedly asks, 'How random is random? You know more than you think. You know where you cut in' (Burroughs and Gysin 89). Joyce would probably concur. For the editor – the secondary 'scissors and paste man' – to focus on the vocabulary of the conscious, or to seek to correct certain distortions, is to perform a sadly normalizing operation. George Steiner claims: 'He who passes over printing errors without correcting them is no mere philistine: he is a perjurer of spirit and sense. It may well be that in a secular culture the best way to define a condition of grace is to say that it is one in which one leaves uncorrected neither literal nor substantive errata in the texts one reads and hands on to those who come after us. If God, as Aby Warburg affirmed, "lies in the detail," faith lies in the correction of misprints' (*No Passion Spent* 7). This injunction, like so many of Steiner's, bespeaks a fierce classical humanism. Such a claim may be valid in relation to stabilized, canonically endorsed works. I have my doubts, however: whatever the status of the 'secular culture,' its imperative mode is abandonment of absolutes, not adoption of half-measures.[2] (Think of the debased 'Grace' effected in knocking one's hat back into shape, earning time to pay one's gambling debts by materialistically 'wash[ing] the pot' [D 162], in lieu of correcting the soul.) In any case, the faithful act of 'the correction of misprints' certainly does not apply to *Finnegans Wake*, where, Steiner himself admits in another context, the 'fissure opens' (*After Babel* 189). If the 'main task of the editor is to eliminate error,' argues Mahaffey, then the process 'literally carried out in the editing of Joyce would obscure his *modus operandi* in *Ulysses* and eliminate *Finnegans Wake* almost entirely' ('Intentional Error' 183). Luckily this has not yet occurred, but I will turn to developments in editing Joyce, below. For now, the general relation of editing and error, as an aesthetic-hermeneutic problem, remains the untouchable subject in much of textual theory. In a critical dialogue on reading, editing, and critical dialogues, McGann splinters possible perspectives on editorial responses to texts, at one point asking: 'What is the status of error, evil, failure in poetical work? ... most are happy to imagine the carnival of interpretation, the dialogue of endless errant reading. But if the primary texts are themselves errant and ideological, how are we to read them? Certainly not as transcendent models. They seem, in this view, more like images of ourselves: confused, mistaken, wrong – and perhaps most so when we imagine them (or ourselves) reasonably clear and correct' (*Black Riders* 158). I will revisit 'the carnival of interpretation' in a later chapter; here, I wish to retain focus on the 'models' and how they are made.

Editing as a single verb can be misleading. Three formal concerns can be seen to constitute the act of creating and publishing an edition, though none of them is completely separable in consideration from the other two. These three aspects of an edition can be termed sequence, presentation, and annotation. I will briefly describe the crises they represent in general before freely using them in consideration of editions of Joyce.

The first of these aspects is the most readily understood, since it is too often thought to be the entire responsibility of editing. Text^3 needs to be spelt out correctly: the right words in the right order, as it were. Every letter and point of punctuation is a character in sequential relation to every other character. Sequencing as editorial act has now significantly broken from its bonds in the Gutenberg Galaxy (though in an oral form it was always free of them); the sequence of text can be considered as a virtual problem, since one may scroll through computer files full of characters but not be actually 'committed' to paper. The most contested procedure in sequencing is, obviously, the establishment of a copy-text or other kind of principal source. Recent initiatives in editing and textual studies have challenged the need for this source to be anything but virtual – but I will return to this issue in consideration of *Ulysses*.

By presentation, the second fact of an edition, I mean the physical manifestation of the text. Layout, font, paper stock, ink colour, book dimensions, cover design, number of volumes, and related qualities all are matters of presentation. Contrary to 'can't judge a book' clichés, these facets of a book are inextricably linked to a reader's acts of interpretation, and the sensory information one absorbs, consciously or not, is hardly limited to the shapes of marks on the pages, or even to the visual domain itself. The weight, feel, and even the smell of the book (think of the 'almost no smell' of the flower Martha includes in her letter to Flower/Bloom [*U* 95]) are far from irrelevant to the cognitive feat of reading. In *A Rationale of Textual Criticism* G. Thomas Tanselle writes that editors must 'realize that what they are attempting cannot be fully achieved: if the interpretation of a text depends in part on the physical evidence of the document in which the text appears, they must perforce deprive their readers of that evidence, for whatever kind of reproduction they make cannot be the same physical document' (58). Tanselle's point here is strong even as the age of mechanical reproduction is transmogrified into the age of virtual simulation. Implicit in these concerns of reproduction is the physicality of the work: the original and/or the edition. Of the three facets of editing being outlined here, this one is probably the least appreciated, particularly in the age of hypertext transitions (again, more on this point shortly).

Annotation, finally, is the declared critical interaction between edition and edited text. Every preface, footnote, glossary, or index – Gérard Genette has termed these features 'paratexts' – represents an attempt to explicate not only the edited text, but the modus operandi of the edition itself. So, for example, a marginal gloss defining an archaic or a foreign word or an expression in a text remarks upon the cultural separation between the author of the text and its editor. The central problem of annotation (most obvious in complex texts, such as those by Joyce) concerns its own limits: what in the text warrants how much commentary? Sequence and presentation pay the costs of annotation's excesses. In Buck Mulligan's skewering of faddishly parochial Celticism there is the note of Joyce's prescience of how his own writings will be nearly overshadowed by the industry of explication surrounding them: 'That's folk, he said very earnestly, for your book, Haines. Five lines of text and ten pages of notes about the folk and the fishgods of Dundrum. Printed by the weird sisters in the year of the big wind' (*U* 14). The awareness of this phenomenon is most apparent and responsive in the second chapter of the second book of the *Wake*, in which the proliferation of unhelpful notes surrounds the text, exceeding the self-annotative impulses of *The Tale of a Tub* and *The Waste Land*.

A page from 'Two Gallants,' taken from the Penguin edition of *Dubliners* (see figure 1), demonstrates how sequence, presentation, and annotation struggle against each other. In this edition annotation is clearly the editor's greatest and most ostentatious concern. Terence Brown's notes appear at the end of the book, while the text of the story is cluttered with superscript numbers. Of the eleven notes on this page alone – the story is given seventy in total – four could have been made unnecessary if Brown had chosen to include a map of Dublin in the edition. The reader is not to forget to consult and appreciate the notes,

She's a fine decent tart, he said, with appreciation; that's what she is.

They walked along Nassau Street[27] and then turned into Kildare Street.[28] Not far from the porch of the club[29] a harpist stood in the roadway, playing to a little ring of listeners. He plucked at the wires heedlessly, glancing quickly from time to time at the face of each new-comer and from time to time, wearily also, at the sky. His harp[30] too, heedless that her coverings had fallen about her knees,[31] seemed weary alike of the eyes of strangers[32] and of her master's hands. One hand played in the bass the melody of *Silent, O Moyle*,[33] while the other hand careered in the treble after each group of notes. The notes of the air throbbed deep and full.

The two young men walked up the street without speaking, the mournful music following them. When they reached Stephen's Green[34] they crossed the road. Here the noise of trams, the lights and the crowd released them from their silence.

—There she is! said Corley.

At the corner of Hume Street[35] a young woman was standing. She wore a blue dress and a white sailor hat.[36] She stood on the curbstone, swinging a sunshade in one hand. Lenehan grew lively.

—Let's have a squint at her, Corley, he said.

Corley glanced sideways at his friend and an unpleasant grin appeared on his face.

—Are you trying to get inside me?[37] he asked.

—Damn it! said Lenehan boldly, I don't want an introduction. All I want is to have a look at her. I'm not going to eat her.

—O ... A look at her? said Corley, more amiably. Well ... I'll tell you what. I'll go over and talk to her and you can pass by.

—Right! said Lenehan.

48

Figure 1   *Dubliners* (Penguin) 48. Font: 12 point Monophoto Sabon

even concerning as innocuous a word as 'strangers': 'Traditional mode of reference to the English invasion and occupation of Ireland' (*D* 262). There is a nascent argument, if not a directed reading of the entire story, in this little note. (I take it that this tendency is sanctioned by the inclusion, among the book's publication data, of the unctuous phrase, 'The moral right of the editor has been asserted.') But the scrupulous meanness in the efforts to annotate flaws the presentation of the text and is not present in proofing the sequence: a tiret is missing from Corley's remark at the top of the page.[4]

There are, however, always worse editions when it comes to Joyce. Andrew Goodwyn's edition of *Dubliners*, produced for the Cambridge Literature series, veers away from the tiret altogether: '"She's a fine decent tart," he said, with appreciation; "that's what she is"' (*D-C* 49). Why the radical repunctuation, so clearly contrary to the author's stated wishes and every other edition of the text? Goodwyn offers no explanation in any of his pages of introduction, resource notes, and glossary, but given the pedagogical emphasis of the edition, one is left to suppose that the tiret was feared to be altogether too unfathomable for students.[5] The example of *Dubliners*, that favourite of the classroom, clarifies a simple point. Editions of literary works weigh whatever 'fickling intentions' (*FW* 439.01–2) – the phrase encompasses the various appropriate meanings of 'fickle': deceitful and treacherous, changeable, to flatter, to puzzle – the author may have (had) in composing the text against those active in the editing.

Of course, *Dubliners* does not offer the challenges inherent in Joyce's later texts, where an editor will scratch his or her head for years at 'how minney combinaisies and permutandies can be played' (*FW* 284.12–13). These works, as I have suggested in an earlier chapter, turn out to be as multiform, changing, and shifting, as it were letter by letter, as the distortive language game played at the end of 'Ithaca': 'Sinbad the Sailor and Tinbad the Tailor and Jinbad the Jailer and Whinbad the Whaler and Ninbad the Nailer and Finbad the Failer and Binbad the Bailer and Pinbad the Pailer and Minbad the Mailer and Hinbad the Hailer and Rinbad the Railer and Dinbad the Kailer and Vinbad the Quailer and Linbad the Yailer and Xinbad the Pthailer' (*U* 871). Consistencies, which Robert M. Adams looks for in *Surface and Symbol*, are actually tricks, sleights of hand, and are not for Joyce the essential stuff – the 'erroroots' (*FW* 285.12–13) – of life, or of art. In his 1912 Trieste lecture on Daniel Defoe, Joyce expresses a startling conception of 'the new realism': 'Pedants strained to expose the paltry errors into which the great precursor of the realist movement had fallen. How could Crusoe stuff his pockets with biscuits if he had stripped before swimming from the beach to the stranded vessel? How could he see the goat's eyes in the pitch-dark cave? How could the Spaniards give Friday's father an agreement in writing when they had neither ink nor pen? Are there any bears or not on the West Indian islands? And so on. The pedants are right: the errors are there; but the broad river of the new realism carries them off majestically like bushes and reeds uprooted by the flood' (qtd in Herring 140). (It is easy to see in this early image of the flood the coming 'riverrun' of Anna Livia.) Thus, when Adams explains that 'Joyce was less concerned with intellectual precision than with the machinery of precision, with the click and glitter of accuracy' (182), he misses the point. Shaking his head at the 'arithmetical errors of primitive simplicity' that he finds in the 'Ithaca' comparisons between the ages of Stephen and Bloom (*U* 794), Adams expresses disappointment: 'No doubt the whole idea is comic in its intent; on the other hand, this sort of projection can be seen as a deeply considered device of Joycean perspective, and one would not be shocked to find Mr. Hugh Kenner hailing it as a significant *gnomon*. But if it is a gnomon, the lines of projection are skewed and inaccurate; so it is a warped, fantastic gnomon; and one must decide whether the errors arose from incapacity or intent; and there are very good grounds for determining the matter either way; and neither decision yields us a markedly superior novel. So that in effect Joyce has made more trouble for his reader by being inaccurate than he can ever hope for the passage to redeem' (183). Joyce's argument concerning Defoe's 'errors' alerts the reader to the complexities of the aesthetic entire, while Adams seems to give no thought here to context (the onus to 'redeem' is on the excerpt). The complexities of life (such as muddling a few figures in one's thoughts just prior to retiring to bed after a long and active day) are in the disorienting effects of error, inconsistency, ruptures in pattern. 'Tinbad the Tailor' adapts 'Sinbad the Sailor' by merely changing both of the capital letters to the same letter, retaining a dictionary-confirmed solidity for the noun denoting the occupation of 'Tinbad,' and 'T' follows 'S' in alphabetic order. From these patterns alone one might logically expect the next entry in the sequence to begin with a 'U' (which would violate the rules of the second pattern). Instead, 'Jinbad the Jailer' suggests the association between entries is homophonic with the provisions that the chosen first letters for the nouns be the same and the last word still be a recognizable occupation. 'Dinbad the Kailer' thus seems even more erroneous.

The rules of the text keep changing. Readers of Joyce may take this hint: Sinbad, another form of Ulysses, is a hero and an epic by any misspelling.

Fritz Senn refers to the act of 'righting,' a term borrowed from Bloom's 'righting her breakfast things on the humpy tray' (*U* 65). This verb, Senn writes, is 'convenient shorthand for at least four interconnected processes: (a) characters in the book, mainly Bloom, amending their practices or conjectures in what they momentarily believe to be improvements; (b) Joyce revising and retouching his own handiwork; (c) the reader/critic adjusting to the text; and (d) the book itself tending toward ameliorative diversity' ('Righting *Ulysses*' 12). While '*Ulysses* is probably the first consistently autocorrective work of literature' (Senn, *Joyce's Dislocutions* 69), it 'may also be called, with equal justice, the first self-wronging book' (Senn, 'Righting *Ulysses*' 12). These bizarre qualities are understated by the proliferation of *Ulysses* editions and by the frequently passionate critical responses to them.

In 1984 there was a sea change in editorial practices, with the publication of Hans Walter Gabler's 'Corrected Text' of *Ulysses* (which, notes Michael Groden, 'was the publisher's title and not Gabler's, just as the inane stars at the beginning of each chapter were the publisher's intrusions, but the subtitle nonetheless went out over Gabler's name' [Groden, 'Perplex in the Pen' 234]). Despite Gabler's writing in his Afterword to the edition that 'it should be understood that it is an edited, and not a definitive text. No text written or edited can be wholly divorced from the processes of writing and editing and the decisions and judgements that they entail. Hence, definitive texts do not in truth exist, but at the most approximations to the best possible text' (650), reactions to the book were wide ranging and outspoken. The loudest opposition – there is no contest – continues to come from John Kidd. Heated exchanges between Gabler and Kidd, usually one dismissing the other as incompetent, chiefly constituted the furor known as the 'Joyce Wars' – 'a kind of World Wrestling sideshow for the intellectual crowd,' as Groden has characterized it ('Perplex in the Pen' 235).

The controversy stems from Gabler's postulation of a continuous copy-text rather than a stabilized one. Gabler lays out two guiding principles for editors of texts: 'a general and pervasive one that governs all editing' and another specific, 'all-important' principle concerning *Ulysses*.

> The pervasive principle is that editorial judgment – editorial critical judgment – is integral to any edition. By it, the editor is inextricably bound into

an edition. Or, in more operative terms, the editorial function is a structural dimension of a critical edition. It governs choice and treatment of copytext, recension, and emendation; and it both governs and necessitates the critical apparatus in all its forms, including, in the case of the critical *Ulysses*, the synoptic presentation ... The specific principle for the 1984 edition of *Ulysses* is that, fundamentally, it establishes the text of the work that Joyce (successively) wrote, and not the text manifesting itself through the deviational forces of prepublication transmission at the one particular moment in historical time marked by the work's first publication in book form. ('A Response' 251; ellipsis added)

Even when Gabler's language is not patronizing – he is here responding to one of Kidd's well-publicized cries of *j'accuse* – it is authoritarian and frequently deterministic (witness 'governs and necessitates' and the sour touch of 'fundamentally'). There is here a remarkably different tone from that of earlier writings, such as this article written three years before: '[T]he stability achieved – barring transmissional corruption by which it remains threatened – is strictly that of a specific textual version. It does not cancel out the instability of the text in process, which the author can at most set aside, but never undo. Nor can the editor undo it, and, regardless of the author's attitude, he may choose – indeed, he has the freedom – not to set it aside. Since the instability of the text in process is not cancelled out by the final or any other authorial textual version, it can and should not be editorially neglected – though this is what happens in a critical edition hierarchically oriented towards a stable critical text' (Gabler, 'The Text as Process' 111). That important condition – 'barring transmissional corruption by which it remains threatened' – demonstrates that this stability is not absolute. As Mahaffey notes, 'Gabler is conservative (traditional) in the premium he sets on authorial intention at the expense of accident and circumstance, but radical in his decision to define authorial intention as multiple and changing' ('Intentional Error' 172). The synoptic approach attempts to embrace the inconsistencies that Kidd promises in his ever-forthcoming edition to eradicate (see below, chapter 6). Patrick McGee offers what may the best defence of Gabler's edition – inasmuch as it is a defence, and perhaps because, strictly speaking, it isn't one – when he writes: 'the apparatus of a genetic edition creates the illusion of a uniform textual development leading up to and culminating in the publication of the completed work. It presupposes that the end of the textual development is a stable text. Otherwise, the "genetic" metaphor, which governs the theory of such an edition, makes no sense. The purpose of Gabler's edition is to present the instability of the text in process. Since this is an interpretive act and not simply a reproduction of documents, it is necessary to construct an invariant context against which one can read the history of substantive and accidental textual variation' ('The Error of Theory' 158). The edition is in McGee's view 'a theoretical breakthrough even if it is editorially flawed. It may even be necessary to consider the errors as a byproduct of the theory' ('The Error of Theory' 157). If this second statement is true, then my own criticism of the Gabler edition would run this way: since error's prevalence as trope, sign, and method in the novel and its composition is so central and affirmative, the 'errors' readily attributable to the editor ought to be not a mere 'byproduct of the theory' but clearly indivisible from the entire narrative of the theory and its practice.

This brings me to Danis Rose's recent 'Reader's Edition.' Back-cover blurbs culled from keen *Ulysses* scholars Groden and Senn speak of the edition, respectively, as 'BOLD AND BRILLIANT AND ALSO CONFIDENT AND CONTROVERSIAL' and 'THIS MAY BE THE HANDY, USABLE *ULYSSES* THAT WE HAVE BEEN WAITING FOR.' The capitalization and bright red print are the most enthusiastic features of either of these remarks, since both hit a strong note of impersonal caution.[6] Obviously, the notion of a 'usable *Ulysses*' is somewhat antithetical to my argument here, and in my view anyone holding out for such a thing will find themselves waiting a long time indeed. Roy Gottfried suggests that in Joyce's expression of worry to Sylvia Beach that his 'book will never come out now,' the word 'never' deserves special attention: 'Despite such evidence as various drafts and collated typescripts, each yielding plausible editorial choices the possibility arises there could never be a correct *Ulysses*, because it is governed by deceptive letters twisted and turned in their appearance' (Gottfried 165).[7]

Rose's editorial policy appears to be made of two possibly contradictory principles. The basis of his *Ulysses* is a construct he calls the isotext: 'Inasmuch as it is possible to achieve it, an isotext is an error-free, "naked" transcription of the author's words as written down by him or by a surrogate, positive faults and all, with their individual diachronic interrelationships defined. It is not a transcription, however edited, of any single text, but a blending together of the members of a series or complex of texts' (Rose, 'Introduction' xii). Unfortunately, what these 'positive faults' are is left for the reader to guess. At any rate, Rose's 'blend' is an involved, special preparation of this 'isotext'; a distillation and decoration for a most idiosyncratic dish. The reader of the 'Reader's Edition' is served the text liberally garnished with hyphens (to soften those Joycean compounds) and the dessert of Molly Bloom is sprinkled with the apostrophes and italics she never had (but an 'alternative format,' without the apostrophes, is offered in an appendix for those with a hardier palate).[8] Still, the meal weighs in at 690 pages – less, in fact, than any other edition of the book, in part because it has saved space by reducing the size of those three important letters, S, M, and P. That's still a lot of Irish stew, and one might doubt whether it is realistic to suppose that any slight changes of space or spice (and here I note only the most obvious textual alterations) are going bring many new diners to the table.

No other textual critic has introduced this 'isotext' possibility, and few really understand exactly what its formulation involves. Gabler notes the 'terminological blur[s]' and 'how widely copyreading free play is at work' in the 'Reader's Edition' ('Danis Rose' 566–7), though Mahaffey criticizes a similar (though less boldly confessed) urge for a virginal status of working texts in Gabler, who 'doesn't make a neutral distinction between documents of composition and those of transmission; he ranks them in a way that clearly privileges individual composition as potentially "pure," as opposed to transmitted texts that are presented as less desirable in their adulteration. Such an attitude seems sensible enough in a traditional editorial context, but it stands out conspicuously against the background of *Ulysses*, which takes adultery and adulteration (sexual and verbal "wanderings") as a primary subject and method' (Mahaffey, 'Intentional Error' 180). Gabler might well point out that no 'neutral distinction' can, in fact, be made: editing is as ideologically fuelled as any other interpretive act, and perhaps even more so.

This consideration brings to the surface a question silently buried by the textual historians of *Ulysses*: why is Samuel Roth's pirated edition of the novel not considered at all as a critical edition, however aberrant it may be? After all, the first American edition of the novel (Modern Library, Random House, 1934) was built from Roth's edition – if, indeed, Roth's itself is not the 'first' American edition (what *Finnegans Wake* might call 'a notoriety, a foist edition' [FW 291.27]). Its overriding editing principle is, in a word, Bowdlerization, but then this is a more lucid apparatus than some.

Specific examples speak volumes in these comparisons. Richard E. Madtes cites 'a typically Joycean brand of impishness' in an intriguing instance of the composition of *Ulysses*: 'With an addition to the final page proof Joyce expanded the phrase "a book entitled *Sweets of Sin*" [in the "Ithaca" episode] to "a book of inferior literary style, entitled *Sweets of Sin*" ... At the same time, he crossed out "entitled" and substituted "entituled." A later editor or printer, however, familiar with English orthography but not with Irish humor, failed to appreciate this shift to inferiority, and "corrected" the supposed misprint' (43).[9] This single word, 'entituled,' may be seen as a fascinating forerunner of the *Wake*'s portmanteau, combining as it does 'entitled' with 'titular' (another joke at the 'gentleman of fashion' device), while leaving the association with 'titillate' ready for the licentious reader. This may be read as a 'Freudful mistake' (*FW* 411.35–6) on the part of Bloom, who, after all, is trying to conceal the title of this naughty book from Stephen, and thus as another suggestion that the narrative shape of 'Ithaca' is, strictly speaking, neither objective (entirely distinct from Bloom's wily but weary consciousness) nor scientific (accurate, to the letter). The passage in which 'entituled' appears – or does not appear – merits close attention as a microcosm of the worlds of difference between editing initiatives. Here is the most prevalent ('entitled') version:

> How was a glyphic comparison of the phonic symbols of both languages made in substantiation of the oral comparison? On the penultimate blank page of a book of inferior literary style, entitled *Sweets of Sin* (produced by Bloom and so manipulated that its front cover came in contact with the surface of the table) with a pencil (supplied by Stephen) Stephen wrote the Irish characters for gee, eh, dee, em, simple and modified, and Bloom in turn wrote the Hebrew characters ghimel, aleph, daleth and (in the absence of mem) a substituted goph, explaining their arithmetical values as ordinal and cardinal numbers, videlicet 3, 1, 4 and 100. (*U* 805–6, *U-M* 672, *U-V* 688)

One of the fascinating elements in this scene is the 'absence of mem.' Bloom, who has already shown himself on various occasions to be lax in observing his Judaism, may be making a mistake in his Hebrew alphabet. Bad handwriting might produce ק (usually transcribed in English as qoph) for מ (mem). Now, here is Gabler's version of the passage:

> How was a glyphic comparison of the phonic symbols of both languages made in substantiation of the oral comparison? By juxtaposition. On the

penultimate blank page of a book of inferior literary style, entituled *Sweets of Sin* (produced by Bloom and so manipulated that its front cover came in contact with the surface of the table) with a pencil (supplied by Stephen) Stephen wrote the Irish characters for gee, eh, dee, em, simple and modified, and Bloom in turn wrote the Hebrew characters ghimel, aleph, daleth and (in the absence of mem) a substituted qoph, explaining their arithmetical values as ordinal and cardinal numbers, videlicet 3, 1, 4, and 100. (*U-CE* 563–4)

Note the inclusion of an entirely new sentence ('By juxtaposition') and that 'goph' has become 'qoph.' Gabler's edition stands alone in its decision to place a comma after the number 4 in this last sentence. (The last two features are the only differences from the 1922 first edition [*U-F* 640].) Not counting paragraph indentations and the decision to revert to 'entitled,' Rose's edition has four differences from Gabler's, some of them bizarre:

How was a glyphic comparison of the phonic symbols of both languages made in substantiation of the oral comparison? By juxtaposition. On the penultimate blank page of a book of inferior literary style, entitled *Sweets of Sin* (produced by Bloom and so manipulated that its front cover came in contact with the surface of the table), with a pencil (supplied by Stephen) Stephen wrote the Irish characters for gee, ah, dee, em, simple and modified, and Bloom in turn wrote the Hebrew characters gimel, aleph, daleth and (in the absence of mem) a substituted qoph, explaining their arithmetical values as ordinal and cardinal numbers, *videlicet* 3, 1, 4 and 100. (*U-RE* 600–1)

A comma is introduced before 'with a pencil.' This is an alteration away from manuscript evidence towards normative syntax. The changes to the spelling of letters ('ah' and 'gimel'), as well as the decision to italicize 'videlicet,' seem to be gestures representative of the mediative (literally 'simple and modified') function Rose assumes to perform. It is these gestures that prompt Senn to write candidly of the 'Reader's Edition,' '[t]he *Ulysses* which I have come to like is one that displays a flawed world, characterized by fallibility, where characters misremember, misquote, where Bloom flounders – in other words, a funnier book' (Senn, 'Prodding Nodding Joyce' 582).

New directions into hypertext do not so much introduce new problems as they complicate exponentially the most basic ones, namely, the three unmistakable marks of an edition: sequence, presentation, annotation. At this point Groden's hypertext *Ulysses* (or, as it really promises to be, hypermedia, since photographs, maps, musical recordings, and film clips will be included) can be discussed only as a possibility, for it is very much *in utero* and has a long gestation period ahead. The roughness of this beast waiting to be born is inculcated in its designer's aspirations; given the current stage of technology, it is sensible that this edition is to be primarily a pedagogical aid, a tool for consultation, rather than an artefact to curl up with in an easy chair. Consideration of presentation focuses on the toolbars and icons to be used to (a) move or work within the text (push ahead/back to or highlight a chapter, exact page, word, or phrase), (b) access material supplementary to the text (the other media listed above), and (c) cooperate with the text as instantaneous annotation (the perpetual image of a clock, for example, may tell the reader the time of the events described on any given page). Annotative anxieties, given the removal of the reassuring constraint of so many possible pages between covers, require a host of scholarly editors for the project, who are sure to argue among themselves about the limitations of their efforts.[10] Gabler debates notwithstanding, whether the theory and practice of editing as it has been hitherto understood can produce at any rate that is reasonably comparable to the expansions of media and communication technology is the question Groden and his assistants – and textual editors everywhere – now face. To do so, even to attempt to do so, will require a new flexibility not entirely corralled by inhibitive rectitude. The devil of Error will need more dues.

To date there has been no editorial flurry surrounding *Finnegans Wake* comparable to that of *Ulysses* – surely largely because fewer scholars bother to read the *Wake* – but with the outbreak of genetic enquiry, the publication of the *Wake* notebooks in Buffalo, and the forthcoming four-part critical edition by Rose, that may be about to change. (There are hypertext forms of the *Wake* in the works, too, but none yet so ambitious as Groden's *Ulysses* project.) There is no 'best authenticated version' (*FW* 30.10) of the *Wake*, and I think editors are understandably timorous when confronted with the problems such a text flaunts. Senn is worth quoting at length here:

> *Finnegans Wake* defies even more norms [than *Ulysses*], makes it difficult for us to trust any of them or any of the verbal appearances. It amplifies most Ulyssean features and is even more rigorously autocorrective. It too offers serial pluralities of tentative, concealed, often contradictory ac-

counts, refusing even to distinguish among facts, fictions, rumors, myths, fears, and the like, but it goes far beyond successive qualification. *Finnegans Wake* inclines toward instant repair, toward simultaneous retraction, as often as not within one word. Its compressed, fractured language can be seen, from the point of view stressed here, as an attempt to rectify the errors of assertive simplification at once, or to improve on one mistake by interlacing another. Alternative readings are not so much lined up in succession as integrated in the microstructure. The pretense of a simplistic truth is no longer upheld, but yields to a choice of rival improbabilities.

(Joyce's Dislocutions 69–70)

The text of *Finnegans Wake* is, quite literally, an accident waiting to happen. Despite Joyce scholarship's respectful adoption of the universally paginated Viking edition, its pages are filled with strange puzzles of syntax and punctuation. On one page, an end bracket has no corresponding start bracket or even punctuation to mark the end of the sentence – 'you must, how, in undivided reawlity draw the line somewawre)' (292.32) – while on another, a dangling word suggests an entire sentence missing, and 'op. cit.' upset becomes 'opsits':

> And ere he could catch or hook or line to suit their saussyskins, the lumpenpack. Underbund was overaskelled. As
> – Sot! sod the tailors opsits from their gabbalots, change all that whole set. (*FW* 324.12–15)

Though there is much that is fluid in the writing of this book, the estrangement is more than linguistic, and the reader needs to adapt to interruptive discrepancies. Tanselle agrees that '[t]he act of interpreting the work is inseparable from the act of questioning the text' (32), though in the extreme case of the *Wake*, the questions are propagated rather than answered, and even the title of the book (and this holds true, I think, for both 'Work in Progress' and its published title) functions 'to inculcate an awareness that we are all editors, that reading is itself a transitory editorial practice' (Mahaffey, 'Intentional Error' 186). Because the 'writing thithaways end to end and turning, turning and end to end hithaways writing and with lines of litters slittering up and louds of latters slettering down' (*FW* 114.16–18) is continuous and unabated, all editing practices that are not 'transitory' amount to dam-building, exemplified by the bizarre contrivance of Anthony Burgess.* His *Shorter Finnegans Wake* is 374 pages shorter than the full text, and even the pages that remain are not all Joyce's text; for some are occupied by square-bracketed summaries (and more will be said about such unfortunate scaffolding in chapter 7).

Efforts to incorporate within literary anthologies works such as *Ulysses* and, much less commonly tried, *Finnegans Wake*, almost always show strain – not only because the clinical butchery of selection isolates under glass the heart (at best, the heart: or is a kidney more appropriate?) from the circulation system of ever-operative, living context. Just as Blake's works, for example, are purposefully individuated by differences of colour, these works of Joyce are slippery for their continuous self-reference (a 'Tip' from *Ulysses* [375] is always returned in the *Wake* [8.08, 8.11, 8.15, etc.] or vice-versa). These anti-normative impulses constitute a perverse sort of defence gesture, a self-deconstruct mechanism. Inclusion of four pages of *Finnegans Wake* in recent editions of *The Norton Anthology of English Literature* reflects something of a loosening of the bounds on the canon of 'English' modernism, but close study of these pages is revealing. Comparing the Viking edition (figure 2) and the *Norton Anthology* (figure 3), one is immediately struck by the spatial privilege the latter gives its footnotes and their selective nature ('Madammangut!' is passed over, while note 44 glosses a short, relatively uncomplicated sentence [uncomplicated, that is, compared with many others in the *Wake*]). The *Norton*'s editors humbly admit that a 'complete annotation of even this passage is, of course, a physical impossibility *in this anthology*' (2309; emphasis mine). Annotation is represented by these editors as a completable act, but I doubt any serious lexicographer or linguist would take in this claim without blinking.

Again, the cost of the overzealous annotation is paid by the presentation and sequence. The *Norton Anthology*, for its ungainliness, becomes for many students the bane of their bookshelves, and I doubt that, given the choice, a student would prefer to face the page reprinted here from the *Norton*, with its intimidating half-page of notes, rather than the clean Viking page. Perhaps worst of all, the *Norton* has muddled the text: Viking has 'beyond Brendan's herring pool' (213.35–6) where Norton puts 'beyond the Brendan's herring pool,' and in line seven of the same page of *Norton*, 'Las Animals' should be 'Las Animas.'*

These errors are not so unusual but are, in fact, rather representative of an important phenomenon. The *Wake* prevents itself from being enshrined as a mordant museum piece – what anthologies such as the *Norton* effectively do to their constituent selections – and thus remains a loose canon. The exception I find to this trend of poor incorporation –

beads went bobbing till she rounded up lost histereve with a marigold and a cobbler's candle in a side strain of a main drain of a manzinahurries off Bachelor's Walk. But all that's left to the last of the Meaghers in the loup of the years prefixed and between is one kneebuckle and two hooks in the front. Do you tell me that now? I do in troth. Orara por Orbe and poor Las Animas! Ussa, Ulla, we're umbas all! Mezha, didn't you hear it a deluge of times, ufer and ufer, respund to spond? You deed, you deed! I need, I need! It's that irrawaddyng I've stoke in my aars. It all but husheth the lethest zswound. Oronoko! What's your trouble? Is that the great Finnleader himself in his joakimono on his statue riding the high horse there forehengist? Father of Otters, it is himself! Yonne there! Isset that? On Fallareen Common? You're thinking of Astley's Amphitheayter where the bobby restrained you making sugarstuck pouts to the ghostwhite horse of the Peppers. Throw the cobwebs from your eyes, woman, and spread your washing proper! It's well I know your sort of slop. Flap! Ireland sober is Ireland stiff. Lord help you, Maria, full of grease, the load is with me! Your prayers. I sonht zo! Madammangut! Were you lifting your elbow, tell us, glazy cheeks, in Conway's Carrigacurra canteen? Was I what, hobbledyhips? Flop! Your rere gait's creakorheuman bitts your butts disagrees. Amn't I up since the damp dawn, marthared mary allacook, with Corrigan's pulse and varicoarse veins, my pramaxle smashed, Alice Jane in decline and my oneeyed mongrel twice run over, soaking and bleaching boiler rags, and sweating cold, a widow like me, for to deck my tennis champion son, the laundryman with the lavandier flannels? You won your limpopo limp fron the husky hussars when Collars and Cuffs was heir to the town and your slur gave the stink to Carlow. Holy Scamander, I sar it again! Near the golden falls. Icis on us! Seints of light! Zezere! Subdue your noise, you hamble creature! What is it but a blackburry growth or the dwyergray ass them four old codgers owns. Are you meanam Tarpey and Lyons and Gregory? I meyne now, thank all, the four of them, and the roar of them, that draves that stray in the mist and old Johnny MacDougal along with

Figure 2 *Finnegans Wake* (Penguin-Viking) 214. Font: Garamond No. 3

the Brendan's herring pool[23] takes number nine in yangsee's[24] hats. And one of Biddy's[25] beads went bobbing till she rounded up lost histereve[26] with a marigold and a cobbler's candle in a side strain of a main drain of a manzinahurries[27] off Bachelor's Walk. But all that's left to the last of the Meaghers[28] in the loup[29] of the years prefixed and between is one kneebuckle and two hooks in the front. Do you tell me that now? I do in troth. Orara por Orbe and poor Las Animals![30] Ussa, Ulla, we're umbas[31]all! Mezha, didn't you hear it a deluge of times, ufer[32] and ufer, respund to spond?[33] You deed, you deed! I need, I need! It's that irrawaddyng[34] I've stoke in my aars. It all but husheth the lethest zswound. Oronoko![35] What's your trouble? Is that the great Finnleader[36] himself in his joakimono[37] on his statue riding the high horse there forehengist?[38] Father of Otters,[39] it is himself! Yonne there! Isset that? On Fallareen Common? You're thinking of Astley's Amphitheayter where the bobby restrained you making sugarstuck pouts to the ghostwhite horse of the Peppers.[40] Throw the cobwebs from your eyes, woman, and spread your washing proper! It's well I know your sort of slop. Flap! Ireland sober is Ireland stiff.[41] Lord help you, Maria, full of grease, the load is with me! Your prayers. I sonht zo![42] Madammangut! Were you lifting your elbow, tell us, glazy cheeks, in Conway's Carrigacurra canteen? Was I what, hobbledyhips?[43] Flop! Your rere gait's creakorheuman bitts your butts disagrees.[44] Amn't I up since the damp dawn, marthared mary allacook, with Corrigan's pulse and varicoarse veins, my pramaxle smashed, Alice Jane in decline and my oneeyed mongrel twice run over, soaking and bleaching boiler rags, and sweating cold, a widow like me, for to deck my tennis champion son, the laundryman with the lavandier flannels? You won your limpopo[45] limp from

23. The Atlantic Ocean. St. Brendan was an Irish monk who sailed out into the Atlantic to find the terrestrial paradise.
24. Yankees' + Yangtze (river in China). The de Dunnes have swollen heads now that they have emigrated to America.
25. Diminutive form of the name Bridget. St. Brigid (or Bridget) is a patron saint of Ireland. "Biddy" is also a term for an Irish maidservant.
26. Yester eve (last night) + eve of history. The sentence may be paraphrased: "Irish history got lost when she went off in a side branch of the main Roman Catholic church, and Biddy (i.e., Ireland) landed herself in the dirt." There are also Freudian implications here.
27. A urinal + Manzanares (river in Spain).
28. Thomas Francis Meagher, Irish patriot and revolutionary, who was transported to Van Diemen's Land in 1849 and escaped to America in 1852.
29. Loop + *loup* ("wolf") and also "solitary man," French). Cf. Wolfe Tone, the ill-fated Irish revolutionist.
30. Souls (Spanish) + the name of a river in Colorado. *Ora pro nobis* (pray for us; Latin) + Orara (river in New South Wales) + *pro orbe* (for the world; Latin) + Orbe (river in France). The entire sentence may be read: "Pray for us and for all souls."
31. *Umbra* (shade; Latin) + Umba (river in Africa). "Ussa," "Ulla," and "Mezha" are also river names; each contains a number of other meanings.
32. Bank (of river).

33. *Spund* (bung; German).
34. A multiple pun: Irrawady (river in Burma) + irritating + wadding. This and the following sentence may be paraphrased: "It's that wadding I've stuck in my ears. It hushes the least sound."
35. *Oroonoko* (novel by Aphra Behn about a "noble savage," published ca. 1678) + Orinoco (river in Venezuela).
36. Fionn mac Cumhail (Finn MacCool), legendary hero of ancient Ireland.
37. Comic kimono. *Joki* is the Finnish word for river; the name Joachim is perhaps also implied.
38. According to tradition, Hengist was the Jute invader of England (with Horsa), ca. 449; he founded the kingdom of Kent.
39. Father of Waters (i.e., the Mississippi) + Father of Orders (i.e., Saint Patrick).
40. Philip Astley's Royal Amphitheatre was a famous late 18th-century English circus, specializing in trained horses. "Pepper's Ghost" was a popular circus act. One of the washerwomen has been reproving the other, who thought she saw the great Finn himself riding his high horse, by telling her that once before she had to be restrained by a policeman for making "sugarstuck pouts" at a circus horse.
41. The temperance reformer Father Matthew had as his slogan "Ireland sober is Ireland free."
42. I thought so + Izonzo (river in Italy).
43. Hobbledehoy + wobbly hips.
44. The sentence is a punning discussion of her hard work and ailments.
45. A river in south Africa.

Figure 3 *The Norton Anthology of English Literature* (vol. 2) 2311. Font: Bernard Modern 'the one and only time when our copyist seems at least to have grasped the beauty of restraint' (*FW* 121.29–30) – is in *Imagining Language*, edited by Jed Rasula and Steve McCaffery. In this anthology, selections from the *Wake* do appear in columns, bending the layout to suit the book's shape, but there are no logographical distortions. What may be most interesting about this collection's use of the *Wake* is its de-emphasizing of such a work's being representative of an enclosed category, such as 'English literature.' Rather, it seems to me, *Imagining Language* includes its samples from the *Wake* as a localized example of (ongoing) experiments in poetic language.

Roland McHugh sees the idea of producing a 'correct' edition of the *Wake* essentially – to give him the words of the text in question – '[w]ringlings upon wronglings among incomputables about an uncomeoutable' (*FW* 367.31–2): 'No two manuscript specialists can ever be expected to agree on what ought, and what ought not, to be altered. The conception of a 100 per cent accurate text of [*Finnegans Wake*] strikes me as a dangerously idealistic abstraction' (*The* Finnegans Wake *Experience* 79). Yet there would be some justice in applying these last words to the feat of reading the book.

It cannot be said that we have too many editions of Joyce's works, especially *Ulysses* and *Finnegans Wake*, since so many of them are still seeking even first-time readers, and they collectively teach us about Joyce's transfiguration within 'the textual condition.' The co-existence of these many editions – with the understanding, of course, that the more meticulous and thoughtful best make this point – is, in fact, a boon to readers and a direct consequence of the intensification of Joyce's aesthetic of error. Tanselle writes: 'If we care about history ... we should treat every copy of every edition with the respect for physical evidence generally accorded only to books in so-called rare-book collections (institutional and private)' (52). In this 'history' of mistakes grows continually the real meaning of the work.[11]

CHAPTER SIX

(Sic) of irony

The present makes it difficult to examine irony with either concision or conviction. I mean this in both a particular and general sense. In particular, today's zeitgeist (a tired and dislikable word reluctantly re-yoked for short, symbolic duty here) eschews almost any degree of genuflection and wonder, and the almost inevitable tone of public exchange of information or ideas, from the academic essay to the most popular mass media, is one of virtually automated cynicism. As an ironist, Stephen Dedalus has nothing on the youth of the century after him. 'Ironic' has become a byword of our age, either despite or perhaps partly because of its predictably regular misuse by television news anchors and pop music singers.[1] As early as 1974 – long before the supersaturation of irony in works by post-Pynchonian novelists such as David Foster Wallace and Dave Eggers – Wayne Booth lamented that 'irony has come to stand for so many things that we are in danger of losing it as a useful term altogether' (2), and '[l]ike the sublime ... irony has seemed to many to have a life of its own' (175), and in 1978 Philip Howard assembled a list of seven implied meanings for the word 'ironically,' as it opens a sentence:

1. By a tragic coincidence
2. By an exceptional coincidence
3. By a curious coincidence
4. By a coincidence of no importance
5. You and I know, of course, though other less intelligent mortals walk benighted under the midday sun
6. Oddly enough, or it's a rum thing that
7. Oh hell! I have run out of words for starting a sentence with.

(qtd in Enright 139)

The fact of Booth's *A Rhetoric of Irony* – a book that, as Alan Wilde observes, 'is in many ways more a defense of civility than a study of irony' (3) – and its concerns are symptomatic of New Criticism's simultaneous fascination with and barely contained anxiety about irony, the trope that walks like a mode. By virtue of flexibility in his approach, D.C. Muecke's *The Compass of Irony* is perhaps the most exemplary of these studies (notwithstanding the efforts of William Empson and Northrop Frye), but even there the note of defeat is heard, however *sotto voce*: '[n]o classification of irony, no list of all the ironical techniques ever practised, will enable the critic immediately to put a tag on every piece of irony he finds' (41).[2]

In general, the continuousness of 'the present' is an enclosure that theoretically insures against irony. Nobody understands cosmic irony better than Orpheus or Lot's wife. Without distance, temporal and/or spatial, from its object, irony obviously cannot manifest itself. This idea is integrated into the German term for 'the present,' *Gegenwart*, compounded as it is by *gegen* – against, or contrary to – and *Warte*, a clear or lofty point of view. (Hermann Broch entitled the essay that may have saved his life *James Joyce und die Gegenwart*.) These principles of irony are no small part of the legacy modernism accepted from romanticism, and for his part Joyce echoes the problems of relative perspective and concentric ripples of irony in his sly use of the term 'parallax' in *Ulysses*.[3] These problems, and the entire apparatus of romantic and post-romantic irony, are tested most fiercely in *Finnegans Wake*, where, as Stephen Heath observes, 'the time ... will be the "pressant" (FW 221.17), not a simple present but a present pressing on, always already hollowed by the mark of the future; the time of the inscription of traces in the infinite movement from the ones to the others' (52). 'Irony ... has no past,' reflects Kierkegaard, but '[i]nsofar as irony should be so conventional as to accept a past, this past must then be of such a nature that irony can retain its freedom over it, continue to play its pranks on it. It was therefore the mythical aspect of history, saga and fairy-tale, which especially found grace in its eyes' (294). Again in a most unexpected place one encounters a strange but apt description of *Finnegans Wake*'s madness and method.[4]

Are *Ulysses* and *Finnegans Wake* symptomatic of (or even in some measure responsible for) the age of relentless irony we as readers and interpreters now find ourselves in (and of which we may get '(sic!)' and '(sicker!)' [FW 76.07–8]), or do they offer an antidote? I argue: both, or neither. If it is effectively redundant to speak of polysemy in *Finnegans*

Wake (and I believe it is), cogent discussion of its ('use of'?) irony is nearly impossible. In *A Portrait of the Artist as a Young Man* and *Ulysses* there are ghostly instances of hermeneutic uncertainty as well as explicit ones of textual incoherence, instability, incompletion; inversely, in *Finnegans Wake* there are momentary flirtations, limited or restrained gestures, towards determinable meaning. For these reasons discourse about irony and those 'hides and hints and misses in prints' (FW 20.11) must continually retain an 'either/or' position.

Inauthenticity dwells in the details. Nowhere is the struggle between intention (author's or reader's; the latter is sometimes forgotten) and effect more volatile within a text than at the fault-line of error, or the perception of error. The 'typo' is irony's site of opportunity: the potential for textual error is directly proportional to the potential for irony.[5] I want to call into question the aporial relationship between error and irony and the shared problem of how these phenomena occur and/or are signified. What Joyce called in a 1937 letter to Harriet Shaw Weaver 'that marvellous marginal monosyllable "Sic"' (L III 397) has in *Finnegans Wake* connotations of disorder rather than cool editorial amelioration. Many critics of the *Wake* have been afflicted with chronic paraphasia, 'correcting' the text with unliteral persistence, effectively *'smoothing irony over the multinotcheralled infructuosities of his grinner set'* (FW 348.32–3; italics in original). In many cases the text is wiped out by glosses (what we think we know); perhaps no literary work is more misquoted than the *Wake* (much more on this later: see the appendix). Editors of *Ulysses* have no less of a problem claiming certain instances of 'bitched type' (U 751) as intentional and integral by virtue of discernible irony; they seek to find which wrongs are right. The fidelity that Walter Benjamin held to be the yield of literalism is, in the case of Joyce's later works, rarely attempted.

There is just as grand an epistemological headache to be found in the notion of writing of irony in, of, or about a text as of speaking (as Pascal observed) of being (see Eco, *Kant and the Platypus* 9–56). On the bright side – how bright it is I leave the reader to judge – neither the problem of inevitable redundancy in signifying 'being' nor that of inescapable (or at least the inescapable possibility of) irony in signifying 'irony' necessitates consideration of an author. *Pace* Wayne Booth, irony is not definitively dependent upon any projection of authorial intention. On the other hand, irony may very well need or feed off of the *notion* of authority. One reader of this book may look at its title's lack of apostrophe and say, 'how ironic, how clever of Conley,' with whatever degree of enjoyment or distaste: whatever its value, the 'irony' is attributed to 'Conley.' Another reader may certainly see nothing in the title *Joyces Mistakes* but incompetence on the part of the author, and say, 'how ironic that Conley should err in a study of errors.' And what if the first reader should overhear the second's pronouncement, and say, 'how ironic that this insensitive dolt should err about the ironic title of Conley's study of errors' – what then?

Friedrich Schlegel in 'Über die Unverständlichkeit' ('On Incomprehensibility'), probably the most seminal of German Romantic expositions of irony, was clutching his brow with this same headache long before this moment:

> Im allgemeinen ist das wohl die gründlichste Ironie der Ironie, daß man sie doch eben auch überdrüssig wird, wenn sie uns überall und immer wieder geboten wird. Was wir aber hier zunächst unter Ironie der Ironie verstanden wissen wollen, das ensteht auf mehr als einem Wege. Wenn man ohne Ironie von der Ironie redet, wie es soeben der Fall war; wenn man mit Ironie von einer Ironie redet, ohne zu merken, daß man sich zu eben der Zeit in einer andren viel auffallenderen Ironie befindet; wenn man nicht wieder aus der Ironie herauskommen kann, wie es in diesen Versuch über die Unverständlichkeit zu sein scheint; wenn die Ironie Manier wird, und so den Dichter gleichsam wieder ironiert; wenn man Ironie zu einem überflüssigen Taschenbuch versprochen hat, ohne seinen Vorrat vorher zu überschlagen und nun wider Willen Ironie machen muß, wie ein Schauspielkünstler der Leibschmerzen hat; wenn die Ironie wild wird, und sich gar nich mehr regieren läßt.
>
> Welche Götter werden uns von allen diesen Ironien erretten können?

> (In general, perhaps the most fundamental irony of irony is that one soon gets tired of being presented with it all the time. However, what we wish to understand here by irony of irony is something that is created in a variety of ways. When one speaks of irony without using irony, as I just now did; when one speaks of irony with irony, without noticing that at the same time one falls into an even more obvious irony; when one cannot escape from irony, as appears to be happening in this essay on incomprehensibility; when irony becomes a manner, and thus ironizes the writer in turn; when one has promised irony for some superfluous paperback, without looking over his supply and now has to produce irony against his will, like an actor who has a stomachache; when irony grows wild and can no longer be ruled.
>
> What gods will rescue us from all these ironies?
>
> [qtd in Dane 114–5; Dane's translation])

This cry for help from absent or unimaginable gods is reminiscent of those found in the *Wake* – 'O Loud, hear the wee beseech of thees of each of these thy unlitten ones!' (*FW* 259.03–4) – and the trope of falling into irony is also readily congenial to a book 'about' falling down and fallen speech. *Finnegans Wake* can be understood as Schlegel's zenith, Kierkegaard's plague. These thinkers reach for a lifesaver that Joyce does not afford his readers.

Let me turn attention to a pertinent, concrete example of an instance where Joyce's texts trouble a reader's sense of both textual stability and the reader's own competence at reading (anything) unironically. Surely it is a happy accident that 'one of the major paleographical cruxes in all of *Ulysses*,' to borrow John Kidd's phrase (qtd in Bates 43), is found in a chapter known as 'Proteus,' the name of the god who changes shape to avoid direct questions. The instance to which I refer is sufficiently infamous. The problem of Simon Dedalus's telegram ostensibly concerns a single letter:

> Rich booty you brought back; *Le Tutu*, five tattered numbers of *Pantalon Blanc et Culotte Rouge*, a blue French telegram, curiosity to show:
>   –Mother dying come home father.
>   The aunt thinks you killed your mother. That's why she won't.
> (*U* 52, *U-F* 42)

More and more editors and scholars are becoming convinced that 'Mother' should be 'Nother'; that a transmissional distortion has occurred to Joyce's script; that an error has been wrongly righted. Gabler's *Corrected Text* has three points of difference from the above:

> Rich booty you brought back; *Le Tutu*, five tattered numbers of *Pantalon Blanc et Culotte Rouge*; a blue French telegram, curiosity to show:
> –Nother dying come home father.
>   The aunt thinks you killed your mother. That's why she won't.
> (*U-C* 35)

Besides implementing 'Nother,' Gabler does not, as a rule, indent for new paragraphs that begin with a tiret (i.e., Joyce's signal of spoken dialogue: more on this feature below) and places a semicolon after 'Rouge.' Yet without genetic evidence at hand, how can a reader or student assess the word 'Nother' as intentional? And connoting what? The ready answer is that 'Nother' is a self-aware error, a manifest irony. This irony may be signalling the sloppy state of telegraphy, bureaucratic bungling, or at least the lack of familial care associated with Simon Dedalus – in any case it concerns a kind of literal infidelity. (From this perspective, the history of this *Ulysses* passage reverberates with further irony, as we will see in the inevitable differences each edition contains.) For most critics 'Nother' is naturally the juicier possibility, since its strangeness seems to demand creative explication. Patrick McGee, for example, posits this 'scribal lapsus' as a 'general statement on the human condition: "Another dying. Father, come home["]' (*Paperspace* 58). Psychoanalytic, feminist, and postcolonial analyses will find fertile soil here.

For more features than just this one slippery letter this passage is a remarkable one, full as it is of the 'hides and hints and misses in prints' of which *Finnegans Wake* speaks, not all of them necessarily operating in smooth conjunction.[6] Context –like beauty, where you find it – for the contentious M/N point bizarrely plays with the problem more than it resolves it. The phrase 'curiosity to show' seems to corroborate the 'Nother' reading, or, in other words, corral the unstable telegram text within the stable and stabilizing metatext of 'Proteus.' Yet there is also that odd use of the tiret, Joyce's own rich booty brought back from his French reading, which does not accompany any of the other written texts that characters in *Ulysses* encounter in the course of the day. That single line gives the telegram its own speaking voice; its presence is a signal of difference. Even the 'echo' of Buck Mulligan that follows the telegram is not marked as an utterance distinct from the narrative and/or Stephen's course of thought. For his 'Reader's Edition' Danis Rose removes the tiret from the telegram, weeds out any possible semicolon, and, perhaps the most unusual decision here, has Mulligan's echo trail off in an ellipsis:

> Rich booty you brought back. *Le Tutu*, five tattered numbers of *Pantalon Blanc et Culotte Rouge*, a blue French telegram, curiosity to show:
>   Mother dying come home father.
>   The aunt thinks you killed your mother. That's why she won't ...
> (*U-RE* 42)

Relevant here is Fritz Senn's modestly worded but telling caution that 'setting the significantly imperfect Joycean universe right in a high-handed act of recreation can only be done at some risk' ('Prodding Nodding Joyce' 580). The risk is that brushing away the palpable possibilities of error also removes the potential for irony and thus produces a staler, less challenging text.

The few explanations I can offer for '–Nother dying' are not entirely convincing but perhaps worth entertaining. Stephen may be speaking to himself, remembering the telegram aloud. Such an action is not impossible to imagine of him whose navel-gazing so guides this episode's language and shifting (though preoccupied) focus that his mind can similarly give voice, as it were, to fanciful encounters. The visit with Uncle Richie is insubstantial, ghostly, maybe a memory but just as likely a speculation, in that troublesome present tense. Yet its featured characters are given the same punctuation signs of speech as the telegram a few pages later – though not always. Richie's voice fades, if you will, from '–Morrow, nephew' to his 'misleading,' then 'dron[ing],' then 'tuneful' whistling, and the insistence that this is '[t]he grandest number, Stephen, in the whole opera. Listen' (*U* 48) has, significantly, no tiret prefacing it. The spectre of the comic uncle, whistled straight out of *Tristram Shandy*, is muted as Stephen's interest in the scene itself fades. Arguing that Stephen 'gives voice' to the telegram thus has some good basis, but the phrase that introduces with a colon the text of the telegram, 'curiosity to show,' then has a somewhat incongruous verb. Stephen does not 'show' it but 'reads' it (and to whom would he show such a telegram anyway?), at least according to this interpretation of the tiret.

If the tiret for Joyce assumes the powers possessed by inverted commas in most English texts, perhaps it can function as a deliberate signal of irony. Henry James is probably the master of the technique of ironic quotation: 'It must be admitted that holding one's self to a belief in Daisy's "innocence" came to seem to Winterbourne more and more a matter of fine-spun gallantry. As I have already had occasion to relate, he was angry at finding himself reduced to chopping logic about this young lady; he was vexed at his want of instinctive certitude as to how far her eccentricities were generic, national, and how far they were personal. From either view of them he had somehow missed her, and now it was too late. She was "carried away" by Mr. Giovanelli (144). No true friend to Winterbourne, the narrator of 'Daisy Miller' has a persistent habit of disowning the language of his narrative. 'James's use of inverted commas,' observes Muecke, 'is only a more overt way of "placing" his subject and distancing himself. The inverted commas deny any connexion between such usages and himself: they are not *his* words' (57; italics in original). In *Joyce, Joyceans, and the Rhetoric of Citation*, Eloise Knowlton writes: '[a]s a means of enacting ironic distancing from the grip of language, the inverted comma constructs a rhetoric whereby the subject stands not only outside THAT language, but outside language itself, entering or retreating "at will["]' (17). Nabokov once quipped that 'reality' was perhaps the one word that could not be rightly expressed without quotation marks around it, but 'irony' may well share this trait. Knowlton herself is forced to use the distancing quotation marks in her statement(s) about them, and Muecke's italics are to the same purpose: irony quickly becomes a tautology of form. Joyce's well-known distaste for 'perverted commas' (*L III* 99) probably prevents him from employing exactly this strategy – at least consciously! – but there are other typographical tricks up his sleeve (consider, as only one example, the sardonic weight of italics in the word *'artistes'* as it appears in *Dubliners*).

Morse code offers another explanation (no pun intended). The difference between the two possible words when encoded is minimal but revealing:

—— ——— — .... . .—.

(MOTHER)

—. ——— — .... . .—.

(NOTHER)

If the dash of '–Nother' is considered part of the text of the telegram itself, or even the mental emendation of Stephen, it constitutes the Morse dash that has been neglected, overwritten by a dot.[7] Giulio de Angelis's Italian translation of the novel adopts the idea of 'Nother' as a surrogate 'Mother' but has to extend the error with a telegraphic stammer:

> Che po' po' di bottino ti sei portato via. *Le Tutu*, cinque numeri sbertucciati di *Pantalon Blanc et Culotte Rouge*, un telegramma francese azzurro una curiosità da far vedere:
> –Manna morente torna a casa papà.
> La zia pensa che tu abbia ucciso tua madre. Per questo non vuole.
> (43)

('Manna,' depending on your view of Joyce's cosmogony, may be a right, happy fault or wrongly providential for poor 'Mamma.') At various points in his lifetime Joyce was exposed to telegraphic flubs, and this incident in *Ulysses* bears comparing with the 15 February 1932 telegram received by Stanislaus Joyce: 'GRANDSON BORU TO DAY NAME STEPHEN JAMES' (*L III* 241).[8] Truth makes fiction's errors seem feasible, for there is greater difference of arrangement in Morse code between N and U than M and N.

_... ___ .__. _.

(BORN)

_... ___ .__. ..._

(BORU)

Since the date of this message falls within the era of 'Work in Progress,' a weak argument could propose that the birth announcement is as much a play on words as 'Nother' in the death announcement (might 'BORU' be a proud, punning grandfather's allusion to Brian Boru, late tenth- and early eleventh-century Irish king and hero?), or even a joke meant to puzzle a dour brother.

Not only is Simon Dedalus's telegram a 'curiosity to show,' it is also 'blue' and 'French' (though apparently English in text). Reading 'Nother' specifically as an ironic comment on French telegraphy should take into account the mention of colours in the passage. Taken together, '*Pantalon Blanc et Culotte Rouge*' and 'blue' comprise the national colours of France, the last signifying *la liberté*, perhaps the freedom to err. Furthermore, it is interesting to note that none of the editions cited here alters the title '*Pantalon Blanc et Culotte Rouge*' to *La vie en culotte*, the title of an actual magazine: John Kidd, chomping at the bit to change Mastiansky to Masliansky and by other like means steer fiction into reality, would probably do so. Kidd explains that '[t]here has never been any family in the history of Jewry anywhere in the world named Mastiansky ... I've checked everything – there are no *Mastianskys* on this planet! The name is *Masliansky*. You see, *Thom's* [*Directory* for 1904] had it so gummed up' (qtd in Bates 44). The editorial logic that thus instates 'Masliansky' appreciates neither the importance of textual mediation in even the most mimetic strains of Joyce's representations of Dublin, nor even the fundamental disunity within written universes, a tradition that stretches at least from Dante to Borges. (Certainly one would face a challenge tracing a 'Dedalus' family in Ireland's history.) For Kidd, in the slip-sliding letters of *Ulysses* there is only error, never irony (of which he, as an editor, might otherwise be a ripe target).

There are many other examples of irony-error vortices in *Ulysses*, each with various expressions of potential irony. Certain ironies are reflexively marked or, to use a serviceable term that nods to the convention of '(sic),' bracketed.[9] Stephen's repetition of his 'reconciliation' formula, for example, is checked by an anxious note to himself: 'Said that' (*U* 249).[10] A more complex case is Stephen's 'Green rag to a bull' (*U* 690), a portrait of himself as viewed by the unappeasable Compton and Carr. Green rags have already been considered much earlier in *Ulysses*, and both Stephen's doings with snot and his anti-imperialist wit are established enough for the reader not to dwell on the chance that he makes a mistake about the colour (the word 'bull,' suggesting England's John Bull, is a support to the political joke). The mark of irony seems to be here, mostly in the form of faith in Stephen's credentials as a poet (i.e., his poetic licence). Yet however fleetingly the question may be entertained, it heralds another. Neither red nor green is a wholly correct answer to the question: What colour of flag irritates a bull? Bulls are colour-blind; it is the motion of the cloth that provokes the animal.

In studying the whole 'Mother'/'Nother' problem, as well as the anxious attempts to discern the 'brackets' of an ironic slip, I cannot help but think of another salient and possibly contiguous instance of a 'gumming up.' Before their swords cross in the play's last act, Hamlet offers an unusual apology to Laertes:

> Sir, in this audience,
> Let my disclaiming from a purposed evil
> Free me so far in your most generous thoughts,
> That I have shot my arrow o'er the house,
> And hurt my brother. (5.2.243–7)

Compositor B, something of a notorious quantity in productions of Elizabethan text, allowed 'brother' in the first folio to appear as 'mother'; the first quarto had it set correctly. Obviously, reading the speech as 'hurt my mother' introduces bizarre new ironies to the last act and the play as a whole, not least because Gertrude is only minutes from being poisoned and Laertes's mother is the only member of his family for whose death Hamlet does not seem to be responsible. The sincerity of Hamlet's apology as a whole is, even with 'brother,' hard to gauge and has been read and played as an ironic gesture, either a feigned madness or a contemptuous display.

Here we can borrow a page from Muecke who, in trying to connect irony with intention, or at least trying to determine whether any such possible connection has value, writes: 'It is as if a man who missed the target should say he aimed to miss. He may have aimed to miss, but how, *post facto*, can he prove it?' (57). Shakespeare and Joyce together are 'the boys and errors of outrager's virtue' (*FW* 434.04–5), arch-archers, as free from the need to 'prove' as they are 'from a purposed evil.' Joyce's 'Nother' or 'Mother' is, in a sense, not 'his' any more than 'mother' is Shakespeare's, but the teasing literal ambiguity is the very co-occurrence of 'Joycean' and 'Shakespearean' that makes sense of Stephen's 'absentminded' fascinations with *Hamlet* and mistakes. Shakespeare's own role in Joyce's 'playguehouse' (*FW* 435.02) is one perpetually skewed with textual distortions and misunderstandings. Mr Deasy's use of *Othello* is registered by Stephen as a contextual slip and thus to a considerable degree is a very marked or 'bracketed' instance of irony, like 'Said that.' Yet as a whole, *Ulysses* does not offer the same clear sense of difference to, for example, Bloom's misquotation of *Hamlet* in 'Lestrygonians':

Hamlet, I am thy father's spirit
Doomed for a certain time to walk the earth. (*U* 192)

Bloom's preference for 'earth' over 'night,' it could be argued, reflects his wandering Jew status; but in any case this irony is less explicit than Deasy's Iago.

For an author like Joyce, or any of the other of the most rigorous modernists, the context of any feature of the text – from an allusive phrase to the most bizarre portmanteau word – extends far beyond the text itself. Milan Kundera, with the hard-nosed manner that is so characteristic of his essays, writes: 'Irony means: none of the assertions found in a novel can be taken by itself, each of them stands in a complex and contradictory juxtaposition with other assertions, other situations, other gestures, other ideas, other events. Only a low reading, twice and many times over, can bring out all the *ironic connections* inside a novel, without which the novel remains uncomprehended' (203). This definition, informed by careful ('low') reading of authors such as Musil and Broch, is strangely both succinct and slightly misguided. How many 'ironic connections' are there in *Ulysses*? How is a reader to know s/he has discovered them all? And does not the satisfaction of comprehension – a term I adopt with some apprehension – shut down the dynamic of assertions within the text? In *The Critical Mythology of Irony*, Joseph A. Dane posits the struggle between Socrates and Alcibiades as a reflection of 'Plato's own exegetes,' a 'struggle of the reader against text': 'Irony, thus, is not this action itself, but rather a word which becomes associated with such a struggle; if not a sign of that struggle, the word at least indicates its existence. When a reader, critic, or exegete speaks of irony, that reader is invoking a struggle between reader and text, and likewise the struggle of the reader, not for understanding, but rather for mastery' (30–1). The trope of struggle for mastery is both appropriate and well known in critical assessments of Joyce, and I will return to it and the matter of the 'uncomprehended' novel in chapter 7. Here, I want to ask a fearful question about the discovery – an interesting verb from Kundera – of 'ironic connections' in a work. Spotting and 'tagging' (recalling Muecke's term) ironies, cancelling connections, sounds a morbid business, even an Oedipal mania, with every reader frothing after the hateful father-Socrates substitute, the text. Can it be that to recognize an irony is to negate it? Paul de Man gives tantalizing support to this possibility in his conclusion to *Allegories of Reading*: 'Irony is no longer a trope but the undoing of the deconstructive allegory of all tropological cognitions, the systematic undoing, in other words, of understanding' (301). Booth calls *Finnegans Wake* 'The Encyclopedia of All Ironic Wisdom' (212); this is a wonderful phrase (similarly, Frye calls it 'the chief ironic epic of our time' [323]), though I am not sure what it means. But perhaps that is the point.

So far I have attempted to demonstrate that there is an inherent and challenging 'either/or' relationship between irony and error, but the point that needs pressing here is that this relationship is not an inflexible opposition but a kind of dialectic of co-extant interpretive frameworks.[11] In Kenneth Burke's terms, this dialectic occurs 'where the agents are in ideation,' a crucial separation from drama, where 'the ideas are in action' (512). In this respect, the idea of ambiguity – etymologically, *wandering around uncertainly* – itself becomes ambiguous: it pleasantly becomes a useful term for this dialectic. Burke plainly expresses why this should be so: 'what we want is not *terms that avoid ambiguity*, but *terms that clearly reveal the strategic spots at which ambiguities necessarily arise*' (xviii; italics in original). Although Linda Hutcheon, for one, feels there is some necessary distinction to be made between ambiguity and irony, in her book *Irony's Edge* she is far from helpful on demarcating such a separation, offering only that 'ambiguity and irony are not the same thing: irony has an edge' (33); the repeated assertions of this 'edge,' however, remain unencumbered by a definition. Although she appears to have some reservations about such an arrangement (63), Hutcheon ultimately cannot relinquish the idea of cultural signifiers (words, linguistic or musical phrases, images, etc.) as merely bipolar, either positively or negatively charged, absolutely unironic or ironic (observe her postulation of a 'real "ironic" meaning' [88]). In the case of language and text, which concerns us here, this means that a single word inevitably has only two possible meanings, or at least two limited, exclusive, and probably oppositional sets of meanings. This position demonstrates not only a poor appreciation of language but also a very limited understanding of language's polymorphous and ever-evolving form. Any page of *Finnegans Wake* ably shows the paucity of such an argument: 'I publicked in my bestback garden for the laetification of siderodromites and to the irony of the stars. You will say it is most unenglish and I shall hope to hear that you will not be wrong about it' (FW 160.20–3). English or unenglish; ironic or unironic; will or will not?

I return to my earlier response: both, or neither. The dynamic is vital; to determine and differentiate is negatory. As Dane remarks, '[w]hether the object of attention is a pun, a slip of the pen, a minor poem, a neglected genre, or even an eighteenth-century misprint – that object can be seen as representative of the entire field of literature. The detail tends toward the universal' (3). Muecke is more expansive on this point: 'There is a potential for irony in the very nature of art if we regard it as aiming at both the particular and the general, as both an activity and the result of an activity, as the product both of conscious planning and unconscious spontaneous invention ... There is also room enough for irony in relationships between the artist and his work ... [in works like Mann's *Tonio Kröger* and *Doctor Faustus*] we find, for example, the idea that the artist must be a spiritual cripple, dehumanized, an *âme damnée*, even satanic, so that his art shall be healthy or human or at least so that it can celebrate the life the artist has to forgo. But conversely, the artist can maintain or restore his own healthy attitudes by writing a "sick" book' (163–4).[12] Unnecessary psychobiographical musings notwithstanding, the idea of 'writing a "sick" book' fits the artist who calls himself 'Katharsis-Purgative' and fills his pages with paralysis, stutters, bile, hoof and mouth and venereal diseases. The contagion that is *Finnegans Wake* leaves every reader hiccuping 'sic' after enunciating or writing out every diseased word (see the *Wake*'s own use of the device: *FW* 76.07–8, 260.R2, 368.14–15) and thus achieves 'the point of highest perfection for Schlegel, that is, of a perfection which is conscious of its own imperfection by inscribing this feature into its own text' (Behler 84), or, correspondingly, what Burke refers to as irony's 'internal fatality' (517). There is no peace from irony within such a volatile text-space, only 'paisibly eirenical' (*FW* 14.30) errors.

Literature as a collective enterprise may well be a body that needs occasional if not frequent harsh inoculations of innovation, anti-tradition, revision – but also irony via instability. Joyce, who graduated from medical student to radical textual surgeon, develops in *Ulysses* and especially in the *Wake* a virus that infects all attempts to write about these works. Joyce forces us to err and, consequently, compels us to be ironic about it. Every time we claim that *Finnegans Wake* 'means' something, we are victims to the book's greatest irony; every 'reconstruction' of the ruins/runes is a mistake, a possibly disastrous watering down of a necessary poison.

Like Beckett after Joyce, however, we cannot not write after or about Joyce. Even at the heights of his ironies, Cervantes does not fully reject the value of the impossible quest: so, too, Joyce. Irony, which Ernst Behler contends 'is inseparable from the evolution of the modern consciousness' (73), has a viral existence that requires narrative, even if this narrative takes the form of the modest, errant palliative of criticism. Fritz Senn, who pretends to live happily in Zurich but is truly a citizen of Crete, has cunningly observed: 'All Notes are liars – useful, incomplete, overdone, misconceived, partly irrelevant, and unseasonable liars. The previous sentence is a note. So is the whole of this essay' (*Inductive Scrutinies* 153). I retailor this smooth fabric of thought for my present suit; every word of, on, or about Joyce is ironic in that it wrongs every other word of, on, or about Joyce. In his *Investigations* Wittgenstein remarks, 'Der Philosoph behandelt eine Frage; wie eine Krankheit' (91; 'The philosopher's treatment of a question is like the treatment of a disease'). But a completely healthy doctor is never to be trusted.

Intermittences of sullemn fulminance

> Those quiet cold fingers have touched the pages, foul and fair, on which my shame will glow for ever. Quiet and cold and pure fingers. Have they never erred?
>
> (GJ 13)

Temporality and text have a bumpy relationship: this is something suggested by the preceding chapters (Part I as a whole, in fact), and it is the notion of 'now' or 'the present' or being 'modern' which is specifically the trouble. The interloopings which follow represent a meditative experiment of sorts, in which this text itself is both the subject and the analyst.

It is a sumnny afternoon in April – yes, *that* month – and I am ignoring the flashes of red underkline which appear when I hit the space bar after finishing typing a word: this, obviously, is the cue for the human to notice what is patently obvious to the computer. But like I said, in this experiment, I am paying no mind to the spellchecker's naggings, and will push on. Also I refuse to use any delete function, or to revise in any fashion what I am about to write (or am writing: the difficulty of writing about the tense of an argument in text while in the textual argument oneself!), although I do look at the screen to follow the path I am hacking out for myself. Finally, my last restraint is linear sequentiality to the writing (if not the argument!) , which is not something I much enjoy, preferring by habit to jump back and forth in the manner of reading encouraged by the works of authors like Sterne, Joyce, and Cvortazar.

Let me be clear: I am trying *not* to make mistakes. Distractions have been minimalized – even to the point of inspirational initation, I am wearing a blank white shirt as Joyce liked to do when strecthed out to compose – but of course there is some agitation, which I hope will pass or fade somewhat as I go, at glancing up to see the typographic flubs, such as the injustice of getting poor Cortazar's name wrong. This not exactly automatic writing, and especially not a particularly Dadaist maneuver, but there are resemblances. This deviation from academic rubric is not meant to nettle certain readers, though it probably will. Or does.

When considering how error occurs (this is a loose verb: it tries to encompass with its little might all of the conceptions of manifestation, so give it a chance), by which I mean how it cocurs in a text, what Jerome McGann calls the 'textual condition' must be confornted. Actually, in this instance I am not so much confronting it as immersing myself in it; absorbing the heat, to use more McLuhanisms, of the medium in the process of forming a message. *Finnegans Wake* sits nearby, untouchable, giving a look not of reporach but of a daring lerr: go on, go on, I dare you.

It is a sunny afternoon in April. (Ha! Got it right that time.) I am thinking about Joyce.

Those locaters, the dates and city names which appear at the so-called ends oj Joyce's novels, are tricky things. Earlier I discussed the effects of Joyce's titles (how they offer an integral commentary, a 'reading' of the story) and the search for 'signs' of irony. The ultimate 'postmarks' of the novels have, I think, a mediationg connection between these two discourses. Every edition I know of (or remember, or perhaps misremember) places the postmark data in italics. Certainly this is to offset this data from the text of the novel – though how effective it is might be an interesting question: imagine a reader who wonders how s/he got from 1904 to 1922 in the space of a single soloiloquy or just in half a page without any writing on it at all after that yes-yes-yes gramophone winds down – but the method for differentiation is happlessly the same as that used to discern a title from other information (even the name of its author).

Joyce liked postamrks and he liked signatures: they are, after all, the generic signs of a letter. Every signature, I want to suggest, is a stutter. Remember the typo on Joyce's birth certificate: that was only the beginning. Joyce's birthday is the same as that of his last two books (not counting *Pomes Penyeach*, that is), but a birthday is something of a redundancy. As Georges Perec has remarked, all stories are encmpassed or contained within the phrase, 'je suis né(e)', after which all else is extraneous detail. Joyce was born. *Finnegans Wake* was born. Someone is born in *Ulysses*. Narrative beyond this point is already heading to its own end, just as every birth is a first step toward death. When I spoke above of 'restraint' I meant 'contraint', oh, 'constraint' that is, if 'meant' has any meaning to it. Birth is the constraint. The redundancy of adding 'on February 2, 1882' is just that of Joyce' postmarks. (Postmarks, birtharks.)

What do the postmarks do for the reader? Dates and place names scribbled on the backs of photographs are usually reminders of that which is either forgotten or not readily guessed. In this sense these qualifiers are corrective: so just in case anyone though t the *Wake* was written in toto in 1939, or written earlier in Jouyce's lifetime, or even outside of it (redundancy again: we can assume it was not written in, say, 1756, or 1956). More importantly, these inscriptions draw attenmtion to the act, the process, of photography. The title page of a book announces, this is a book, but it ought not to be forgotten either that covers and other such familiar devices can lie (*Ulkysses* was smuggled under separate cover to the United States, appearing as *The Complete Works of Shakespare*). Subtler are the signs which appear in the text, linguistic or spatial markers such as Joyce's postmarks. The text's self-awareness is so aggressive a neurosis that it must seize the attention, in whatevere fashion, of the reader – just as the typing errors and malapropisms here demand notice. In short, the postmarks are not warnings (too late for that) but something akin to apologies for the erroneous nature of the text it reconizges and frames. This isn't 1904, says the *Ulysses* postmark, this is just a story pretending it is 1904. But then again, because the postmarks are textual entities themselves (if I can call them entities), they are erroneous too: it isn't 1922, and this certainly isn't Trieste.

It is a sunny afternoon in April. Actually, I am running out of afternoon so I must race on – 'Excuse bad writing,' says Milly in her postscript, 'Am in a hrry' – never mind that I cannot even get a quotation right. This is now, and I want to retain the production of this writing on the miswritingsd of Joyce as much as I conscientiously can within the 'now' I'm establishing as a leitmotif. The last thing I want is to be ironic. But of course that last sentence, free though it seems to be of typos, is probably the most mistaken of this series of botched expressions. For that matter, is there anything that can be said (and I mean especially written) about Joyce that is not touched by irnoy? I mean irony.

Joyce used to 'forget' who wrote the big novel about Stephen and Bloom, and would tease others about it. His postmarks are reminders of his connection to the book, for himself; perhaps for the reader they are reminders of the same thing, rather than simply the opposite. Or maybe it's less about character (who) than event (how), the word I have been favouring in previous chapters. It is easier to condiiser the postmark of that other big book, the one sitting nearby here but which I am not actually approaching, physicallly at least. The book of forever, it assures sus. This equals roughly seventeen years. Joyce's postmark may diminish himself, the name or function 'Joyce' as creator. Instead of Joyce, we disover that *Finnegans Wake* was written *by* 1922-1939 (inclusive, presumably), not *during.*, as is usaually the way one interprets final dates. The events of those years in those places shaped the work. Joyce is incidental, or accidental, a vehicle.

This is James Joyce speaking. It is a sunny afternoon in April. A man of genius makes no mistakes etc. etc. The facts are well knoiw in this case. I will take legal action. Viz. Calumny, attributions of error and the like. Please see Patrick Kavanagh's 'Who Killed James Joyce?' for the whole sordid story. I remain. Una stretta di mano. JAS. A. JOYCE. P.S. The difference between a lying text and an erring text is a devi;l's proposition.

It is a sunny afternoon in April.

III
Reading Errors

In spite of careful and repeated reading of certain classical passages, aided by a glossary, he had derived imperfect conviction from the text, the answers not bearing on all points.

(*U* 791–2)

CHAPTER SEVEN

Performance Anxieties

And I shall be misunderstord if understood

(FW 163.22)

Preparatory to anything else, I submit Jorge Luis Borges's definition of reading as 'an activity subsequent to writing – more resigned, more civil, more intellectual' ('una actividad posteror a la de escribir: más resignada, más civil, más intellectual' [*Historia universal* 8; translation in *Collected Fictions* 3]) as a succinct expression of my own conviction. Reading as a verb has been treated elastically to denote interpretation of various kinds ('Mission Control, do you read?'; 'I can read you like a book'; 'how one "reads" a painting'), to the diminution of an appreciative understanding of this unique and special activity of interaction with text. This and the two chapters that follow represent an investigation into how error effects reading – and I leave it to my reader to decide whether the last phrase is a mistake. With this focus in mind and *Finnegans Wake* kept as the central text for consideration, in this chapter I ask: Who reads Joyce? In the next I ask: How is Joyce read? Then, ultimately: How does Joyce read us? Though, as these questions are posed, it should be understood that the particularities (peculiarities?) of Joyce are intermingled with, or, better put, reflexive manifestations of, the general tendencies of literature as an enterprise.

As difficult as it is to configure authorial forms of/for Joyce – to answer 'who in hallhagal wrote the durn thing anyhow' (FW 107.36–108.01) – or to produce a workable text of *Ulysses*, conjuring up for scrutiny the reader of Joyce may well prove even more of a challenge. In Part II, I chose to adopt Derrida's recognition of Joyce as an 'event.'

Academic adoption of *Dubliners* and *A Portrait of the Artist as a Young Man* and quarantine of *Ulysses* and *Finnegans Wake* instead address those works as an eventuality, to which their critical equipment will provide an emergency response: IN CASE OF JOYCE BREAK GLASS. What is not readily assimilated into programmatic academic discourse – at the *Wake* Joyce remains 'our greatly misunderstood one' (*FW* 470.01) – is in the meantime monitored carefully. Umberto Eco's lamentation of the canonization of Thomas Aquinas as that philosopher's worst moment, 'the moment when the big arsonist is appointed Fire Chief' (*Travels* 258), portrays a situation that perpetually threatens to apply to Joyce, but the appointment is always deferred, in part because for every accolade afforded *Ulysses* there are more grumbling dismissals of the *Wake*.

Put very crudely, this book is understood to be the epitome of difficult. It would be fatuous, certainly, to deny or even to marginalize the fact that the level of estrangement projected by the text itself is considerable (who *isn't* afraid of *Finnegans Wake*?). In his contribution to *Our Exagmination Round His Factification for Incamination of* Work in Progress, Frank Budgen looks at the issue of 'difficulty,' if you will, from the text's point of view rather than from that of the piteous reader: 'The difficulty of entering into the imaginative world of Work in Progress lies in no unessential obscurity on Joyce's part but in our own atrophied word sense due in large measure to the fact that our sensibilities have been steam-rollered flat by a vast bulk of machine made fiction. The reader is becoming rarer than the writer. The words of dead poets are read and confirmed like the minutes of the previous meeting, with perhaps the dissentient voice of one Scotch shareholder. Taken as read? Agreed. Agreed' (41). 'Taken as read' is precisely what *Finnegans Wake* is not, in any sense of that phrase. Instead, this literary work's primary notoriety is its status as the most prominent exile in the colony of unread books – a status perpetuated by the publishing and pedagogical warning signs that blockade so many points of entry. A very good example of such sign-building (or scaffolding, as I called it in an earlier chapter) is Roland McHugh's *The* Finnegans Wake *Experience*.* McHugh's title alone suggests that there is something unusual about or more involved than simply reading this book; I have yet to come across *The Three Musketeers Experience* or *The Mansfield Park Reading Tour*. Yes, reading the *Wake* is different, but are its readers? McHugh makes a point of explaining how he himself is, at any rate: 'I spent almost three years reading *Finnegans Wake* ... before looking at any kind of critical account. I contrived to retain this innocence until I had formulated a coherent system of interpretation. I was then able to evaluate the guidebooks from a neutral vantage point and elude indoctrination. Of course, I learned valuable things from them, and had often to discard illformed conclusions in consequence. But this seemed a healthy process, although its duration grotesquely exceeded the time any reasonable person would devote to a book. I hardly intend that my present readers should repeat my example, but I feel that the experience qualifies me to introduce [*Finnegans Wake*] to them in a particularly helpful manner' (Finnegans Wake *Experience* 1–2). I have already remarked on the bizarre hypocrisy of this grandstanding gesture – to appoint oneself a reliable guide after eschewing the idea of such a function – but I want to draw attention to some of McHugh's other, equally dissatisfying assumptions. The first is the invocation of the 'reasonable person,' a character understood at least to have only so much time for any given book. Virginia Woolf's 'Common Reader' is the most defined form of this apparition; these constructs are ghost readers, rhetorical constructs that generally serve either as something with which to bludgeon writers or as straw-man targets themselves. (Think of the puzzled gallery-goer before the upside-down painting mentioned in my introduction.) Budgen's point about the decline in reading needs to be taken – a point made at a historical moment, which made it more prescient than similar statements made, usually with more bombast, in the decades following – and set against contrivances such as the 'Common Reader' and even the 'reasonable person.' *Our Exagmination* effectively does this by including, as it were, 'the dissentient voice of one Scotch shareholder,' the self-confessed 'Common Reader,' G.V.L. Slingsby: 'Whether or not the public can ever be trained to absorb this kind of thing seems to me extremely doubtful. The sort of person who will spend time in the exercise of a new set of muscles such, for instance, as for ear wagging, might be interested in developing a new set of brain or reciving cells, always supposing such cells exist' (190). Leaving the question of cognitive derangement by 'reciving cells' for the next two chapters, let us keep to the matter of 'the public' and its possible needs of a book like *Finnegans Wake*. What agency administers this training? There are more active adherents of intentional fallacy than are dreamt of in poststructuralism. Consider Benstock: 'Much happened during the 1930's to prevent Joyce from mapping out the exegetical attack on the bastion he has built; it is obvious that he did not expect to die without providing many further hints and suggestions for understanding *Finnegans Wake* ... It does not seem too soon to predict that *Finnegans Wake* will never be fully read by any reader (no matter how ideal he might otherwise be)' (40–1). The connection offered between the absence of the author and the (projected) absence of readers is distressingly facile, and Richard Rorty would be the first to identify this view as one of weak pragmatism. As the book sinister admits, 'I know it is difficult but when your goche I go dead' (*FW* 251.26–7); the reader is 'left' to his or her own devices. Or so it seems.

The warning signs posted by critics are never as flexible, applicable, or simply as much fun as those Joyce himself introduced. Consider how Joyce keeps incubating within his work prototypes of possible readers, slouching towards actualization, waiting to be born. The examples of bad or error-prone writers provided in *Ulysses* are many – Mr Deasy, Milly, Martha, Rumbold the hangman – and are fairly well documented in critical studies, but perhaps greater attention ought to be afforded the case of Denis Breen, obsessively bad reader. His desperate scuttling-about, 'hugging two heavy tomes to his ribs,' is a prescient caricature, a warning of sorts to determined readers of the novel. Like Breen, a reader may go '[o]ff his chump' (*U* 201) in furiously pursuing a supposed meaning in even two juxtaposed letters, 'U.P.' (let alone, say, a hundred-letter 'word'). Breen's palpable dementia bears fruitful comparison with Rudy Bloom, found reading 'from right to left inaudibly, smiling, kissing the page' (*U* 702), the very picture of an 'ideal reader suffering from an ideal insomnia' (*FW* 120.13). What better 'ideal insomnia' than the thoughtful repose of the afterlife or, as Rudy is more speculation than spectre, an imaginative extension of Bloom, what more ideal *anything* than the ideals of fiction? (In this regard, to relegate the appearance of Rudy at the end of 'Circe' to a mere hallucination – rather than to appreciate it, say, as an epiphanic vision – is also to reject the 'ideal reader' as a red herring, or is a rhetorical sniff of contempt for the reader of limited attention span. It is to miss the promise of renewal offered by the synergy of the book of life and the book of the dead.) In his well-thumped volume of annotations, Gifford notes that the 'question of which sacred book Rudy is reading has been worried to little avail; it could be any Jewish religious text with the name of God in it' (529). I think there is basis for further argument here, and would suggest that the book is *Ulysses* itself. That each of Joyce's consecutive texts has an awareness – and in the instance of *Finnegans Wake*, a muddled one – of its predecessors has been ably demonstrated by many critics, and even premonitions of or gestures towards the successive text are easily found. Less examined, though I feel more scintillating, is the notion that each text develops a sense of connective exteriority, the *hors-texte* denied by Derrida, to its own textual matter. Put more plainly, Joyce's consecutive texts become progressively more contingent upon themselves as texts. *Dubliners* ends with a tale of a writer presented with an antecedent, external story to his own, and he is haunted by it: 'Other forms were near. His soul had approached that region where dwell the vast hosts of the dead. He was conscious of, but could not apprehend, their wayward and flickering existence. His own identity was fading out into a grey impalpable world: the solid world itself which these dead had one time reared and lived in was dissolving and dwindling' (*D* 225). Gabriel learns that there have been stories other than his own, and he deduces that these stories, which do not run concurrent to his own, must have ended. In a sense he syllogizes: Gabriel is a fiction. Fictions are mortal. Therefore, Gabriel is mortal. There his story ends, but this is not at all what happens to Stephen Dedalus, the would-be writer whose diary entries unexpectedly emerge as narrative device at his *Portrait*'s end. The text literally becomes his, but *Ulysses* is too vast for him, or any other single character in it, to contain. It reads itself, turning back the clock on occasion to revisit another scene, another consciousness, while giving characters inexplicable access to words and ideas that do not cross their paths (Bloom on the 'Rose of Castille,' for example).[1] *Finnegans Wake*, finally, gets ahead of itself by writing and curiously reading its reader – but now I am getting ahead of myself, since this is matter for a later chapter – and refuses to end. The text indicts itself and decrees that it will be 'sentenced to be nuzzled over a full trillion times for ever and a night till his noodle sink or swim by that ideal reader suffering from an ideal insomnia' (*FW* 120.12–14), perhaps the book's most famous phrase. Joyce's writers and writing probe the exterior of the text in which they exist for the reader, seeking to incorporate him/her.

Naturally, the qualities that constitute this 'ideal reader' have been the favourite subject not just of many writings on Joyce or even literature in general, but of studies in aesthetics, semiotics, cognitive studies, and computer programming. Eco has written much on this problem, and he deals with it most cautiously in *The Limits of Interpretation*, where he seeks to balance the role of the text with the role of the reader in interpretation. He suggests a polyvalency to a 'Model Reader': 'when I say that every text designs its own Model Reader, I am in fact implying that many texts aim at producing *two* Model Readers, a first level, or a naive one, supposed to understand semantically what the text says, and a second level, or critical one, supposed to appreciate the way in which the text says so (*Limits of Interpretation* 55; italics in original). This is an intriguing distinction, however simplistic it may seem. In chapter 8 I will deal more with manifestations of the multiple reader; for now the problem I wish to address is that of the psychomachiac pressure the text exerts on the reader if what Eco posits can be considered plausible. Not only does *Finnegans Wake* – and many other texts, according to Eco – demand the powers of the ideal/critical/semiotic reader, it also asks for the attention of the less-than-ideal/naive/semantic reader: 'Come on, ordinary man with that large big nonobli head, and that blanko berbecked fischial eksprezzion' (*FW* 64.30–1). The rules of drama require that a mystery have a Watson for a Holmes, and mystery is the genre Eco uses to make his case for the two-in-one Model Reader principle. As playful and pyrotechnically deceitful as the mystery story may be, however, its genre, most firmly of all genres in fiction, irreversibly predicates a determinable, 'correct' reading: the answer to *whodunit*.

If the text 'designs its own Model Reader,' does it necessarily share or express these designs? Characterizing any text as exclusive in its design of a reader is to refute the experiences of a reader external to the text. When it comes to the case of an author like Joyce, part of this 'experience' involves, with possibly very rare exceptions (but even they are doubtful), clambering through the scaffolding of commentary. McHugh's quest to 'elude indoctrination' is futile, not to mention in bad faith, since it would be rude even to try to come to any sort of *Wake* empty handed. The guides and skeleton keys themselves point to other texts that collectively form contexts for the ominous subject text. Not only are prejudices brought to texts, they are insidiously assigned to them. John Bishop, in one of the best books written on the *Wake*, exposes one of his own prejudices in his introduction: 'It seems to me impossible for any reader seriously interested in coming to terms with *Finnegans Wake* to ignore *The Interpretation of Dreams*, which broke the ground that Joyce would reconstruct in his own "intrepidation of dreams" and, arguably, made *Finnegans Wake* possible ... [Freud's is] an indispensable text to bring to *Finnegans Wake*' (Bishop 16).* *Finnegans Wake* is popularly understood to be madly intertextual, and every critic and key-forger perhaps necessarily flags certain works as at the very least supplementary and most fiercely as 'indispensable' and imperatively prefatory. Why, if the *Wake* does connect with so very many other texts, there should be any evident hierarchy of supplements is not altogether clear. The *Wake* plays upon the neuroses of any reader, no matter the wide reading experience that may be under his or her belt. The devoted student of Bruno, Vico – and/or, for that matter, Freud – is just as tormented as a reader of no acquaintance with those thinkers by the distortive mechanisms of a book that remains, punningly, 'above your understandings' (*FW* 152.04–5). Insofar as the critical scaffolding continually being assembled around the text isolates and barricades it, the specialized *Wake* guides effectively keep trespassers at bay by launching heavy bibliographies at them, lists of necessary preliminary texts and contexts, often keeping silent on the fact that even the most extensive critical incursions have by no means ceased identifying allusions, references, and sources.[2] For these reasons, as well as the inevitable fact of simply human limitations, to speak of ignoring a text like Freud's in the context of the *Wake* is fatuous. *Finnegans Wake* dares us to ignore or dispense with *anything*.

Bishop's phrase, 'seriously interested in coming to terms with *Finnegans Wake*,' is an unusual one on two counts. Reading a book like the *Wake* 'seriously': I wonder if this can actually be done. Maybe the use of the adverb here reflects a disdain for dilettante gestures – both of Joyce's last titles are, sadly, treated by those who have had no experience with them (and even by some who have) more as signifiers of intellectual credentials than as stirring works of art – but it seems conspicuously incongruous with so raucously comic a book; imagine someone claiming to have 'looked seriously at Rabelais.' Is 'coming to terms with' a euphemism, or should it be taken as synonymous with 'reading'? The *Wake* is, before everything and after all, 'where terms begin' (*FW* 452.22). Bishop's phrase is more applicable because it is more direct than many others used to the same purpose (McHugh's title again fits here), but it still feels somewhat unsatisfactory. In his contribution to *Our Exagmination Round His Factification for Incamination of* Work in Progress, Beckett remarks how it is 'inadequate to speak fo [*sic*] "reading" Work in Progress' (15). The 'fo,' faux-'of,' manages to express the inadequacy with fullest impact. To speak fo 'reading' Joyce is not merely naive, as Derrida suggests, but *mistaken*. The fallen language of the *Wake* flows against and over the makeshift dams of 'corrective' language, and ultimately undoes them by provoking errors: 'reversing the findings of the lower correctional' (*FW* 575.33–4).

In many ways *Our Exagmination* continues to be the best volume of *Wake* criticism.* As a collective whole it makes none of the noxious assumptions about Joyce's book or its readership to be found, sometimes only implied, in many later writings. The lack of consensus and the vicariously enjoyable degree of bafflement among the contributors prevent them from stating, indeed, that *anyone* can, should, or want even to attempt to read whatever this thing Joyce is putting together actually is (Victor Llona's essay is blissfully titled, 'I Dont Know What to Call It But Its Mighty Unlike Prose'). Joyce's critics in this volume have their sights set on other critics, too, best exemplified in William Carlos Williams's rejection of Rebecca West; in her dismissal of Joyce, she shows that she 'can only acknowledge genius and defect, she cannot acknowledge an essential relationship between the genius and the defect' (Williams 178). Furthermore, the responsive use of *Our Exagmination* within the *Wake* – various metonymic nods in the direction of its title and the voices of a dozen irreverent disciples, all of whom are also betrayers – reveals an awareness of the text's production within its production and constitutes a remarkable, perhaps entirely unique dialogue between critics and text.

Of course, not even *Our Exagmination* has what McHugh calls 'a neutral vantage point,' but neither does it claim to have one. Those readers of 'Work in Progress,' including, but also besides, the contributors to *Our Exagmination*, should not be denied their individual critical agencies and dismissively painted as duped ground soldiers in the imperious author's advance. Consider Jean-Michel Rabaté's proposal that Harriet Shaw Weaver 'not only rendered the writing of the book possible but also provides the paradigm for the ideal reader intended by Joyce. This reader is a genetic reader, a reader who uses the notebooks and the drafts of *Finnegans Wake*' ('Pound, Joyce and Eco' 488). Weaver's being singled out here strikes me as somewhat peculiar – why not those who took the dictation, or those who scanned the proofs, or even those who translated the book into French under Joyce's supervision? – but Rabaté's overall point about a 'genetic' reader is the salient one. Reference to pre-publication materials is, it might be objected, unfair to the text of the *Wake*, but given the *Wake*'s obsessive interests in intertextuality and its own composition, this objection has little weight. What genetic investigation must not do, however, is assume a role as unchallenged source for verification of readings by authorial reconstitution. A. Walton Litz writes very cogently on the matter of approaching manuscript evidence in this case: 'The evidence of an early version is essentially historical (like the evidence of Joyce's reading or his personal experience): it shows that which is possible or probable, and stands in no absolute relationship to the finished work. But given Joyce's particular methods of word-synthesis and accumulated associations, the degree of probability in the relationship is quite high' ('Uses of the *Finnegans Wake* Manuscripts' 102–3). The stress placed on 'probability' over an 'absolute relationship' makes for a valuable caution, since it necessitates the understanding that Joyce's text is more than a 'singularity' (a term Derek Attridge has recently begun to apply to the *Wake*) because it is not truly a self-contained object or narrative. With this balance in mind, I can here fuse together two of my earlier points: that *Finnegans Wake* overreaches itself, grasps at what seems external to itself, and that the *Wake* rejects no intertextual offer. The *Wake* is a lexical and literary centrifuge, and the 'last word in stolentelling' (*FW* 424.35) is an omega to be perceived but never reached. This, finally, is why 'reading the *Wake*' is an impossible task, or, rather, an unending process.

Critical discourse, especially the kind of discourse I have been calling scaffolding, has a hard time conceding such qualities to any literary work or its interpretation, since to do so would constitute at some level an admission of failure or error into the discourse (something, I have been arguing, that has customarily been anathema to aesthetics and hermeneutics). Instead, then, more or less Procrustean attempts continue to fit a work like *Finnegans Wake* into traditionally recognized parameters for marketing and interpretation. Consider the ways in which the three simple-minded questions below are variously answered.

1. What is *Finnegans Wake*?
2. What is *Finnegans Wake* about?
3. What language is *Finnegans Wake* written in?

The first problem, genre trouble, is one that plagued both *A Portrait* (autobiographical filth) and *Ulysses* (satire on the common man, also filth), at least initially. This is now a question more than sixty years old: 'What is *Finnegans Wake*?' It is not a novel or, at least, it follows none of the traditions of the novel, makes no claims to be a novel, and its author did not refer to it as such. Although Benstock rejects 'mock-epic' as too reductive a classification for the *Wake* (214–15), Margot Norris finds in Joyce's book an anti-novelistic tendency, as 'a critique of the novel itself and, consequently, a critique of the literary and intellectual traditions that have sustained it' (*Decentered Universe* 15), while Sam Slote rightly protests against handling the *Wake* as though it were simply an extension of the novelistic project at work in *Ulysses*: 'Apperceived patterns accumulate and fuse together out of lexical chaos, and thus the critic's task is reduced to an explication of the tension of articulation between tenebrous passages and comprehensible macrotext' (531). Eric McLuhan* sees in Joyce's medley of languages and neologisms, especially in the ten thunderwords, the tradition of Menippean satire, while Donald F. Theall (among others) broadly terms the book 'poetry,' with the implicit understanding that this elastic term will safely repel all objections. 'Even in literature, the highest and most spiritual art,' admits Stephen Dedalus – who does not know the half of it! – in *A Portrait*, 'the forms are often confused' (*P* 214). The lack of consensus here is, I think, another healthy sign of the *Wake*'s resilient immunity to the normative ossification induced by critical commonplaces, the editorial aspects of which I discussed in an earlier chapter. In terms of genre the *Wake* yet remains, as Derek Attridge puts it, 'an unassimilable freak' (Attridge, *Peculiar Language* 10).

McHugh's wish for 'a coherent system of interpretation' takes form in the widespread treasure-hunting reaction to the next two problems. I have repeated a need for suspicion of all statements of what the *Wake* is 'about,' though this is really a necessarily exaggerated form of the stance I would assume towards any other literary text. Here, I will limit this repetition to a look at the stratifying approach to alleged content in the *Wake*. Surveying with suspicion Campbell and Robinson's *Skeleton Key*, Tindall's *Reader's Guide*, and Glasheen's *Census*, Benstock compares the 'titles' given by these authors to the constituent parts of the *Wake* in the mode of chapter titles adopted from Homer for *Ulysses*. I reproduce here Benstock's chart of the four books –

	Campbell and Robinson	Tindall
Book I	The Book of the Parents	The Fall of Man
Book II	The Book of the Sons	Conflict
Book III	The Book of the People	Humanity
Book IV	Recorso	Renewal

– and that of the chapters:

	Campbell and Robinson	Glasheen	Tindall
Ch. 1	Finnegan's Fall	The Wake	The Fall of Man
Ch. 2	H.C.E. – His Agnomen and Reputation	The Ballad	The Cad

Ch. 3	H.C.E. – His Trial and Incarceration	Gossip	Gossip and the Knocking at the Gate
Ch. 4	H.C.E. – His Demise and Resurrection	The Lion	The Trial
Ch. 5	The Manifesto of A.L.P.	The Hen	The Letter
Ch. 6	Riddles – The Personages of the Manifesto	Twelve Questions	The Quiz
Ch. 7	Shem the Penman	Shem the Penman	Shem
Ch. 8	The Washers at the Ford	Anna Livia Plurabelle	A.L.P.
Ch. 9	The Children's Hour	The Mime of Mick, Nick and the Maggies	Children at Play
Ch. 10	The Study Period – Triv and Quad	Lessons	Homework
Ch. 11	Tavernry in Feast	The Tavern	The Tale of a Pub
Ch. 12	Bride-Ship and Gulls	Mamalujo	Tristan
Ch. 13	Shaun before the People	Shaun the Post	Shaun the Post
Ch. 14	Jaun before St. Bride's	Jaun	Jaun's Sermon
Ch. 15	Yawn under Inquest	Yawn	Yawn
Ch. 16	H.C.E. and A.L.P. – Their Bed of Trial	Parents	The Bedroom
Ch. 17	Recorso	Dawn	New Day

(Benstock 5–6)

The differences are telling, whether they are very clearly disparate (such as those of chapters 5 and 12) or apparently slight (chapter 8). Also, the uneven distribution of language and expressions borrowed from the *Wake* is at odds with the use of a word like 'Recorso,' which does not even appear in the *Wake* (or, for the matter, in the *OED*). In the 1999 reissue on the occasion of the book's sixtieth anniversary, readers are provided with an 'Outline of Chapter Contents' – how tellingly elliptical that phrase is! – which is more expansive than the chapter titles listed above only by word count.[3] I reproduce here only a sample, for brevity's sake:

[Book I, Chapter III]
Earwicker's version of the story filmed, televised and broadcast – HCE's wake – Reports of HCE's crime and flight – Court inquiries – HCE reviled – HCE remains silent and sleeps – Finn's resurrection foreshadowed

[Book I, Chapter IV]
Burial of Lough Neugh – Festy King on trial – Freed – Reveals his deception – The letter is called for – ALP is brought in

[Book I, Chapter V]
ALP's mamafesta – The interpretation of the letter – The Book of Kells
(*Finnegans Wake* [1999] xxxi)

The strangeness of these titles and summaries ultimately points to the old adage of tactical training: the map is not the territory. 'We have looked for keys, or else clung to vague analogies,' writes Harry Levin, 'rather than approaching [Joyce] through his boundless particularity' (*Memories of the Moderns* 58). This 'boundless particularity,' which in *Finnegans Wake* is called 'a meticulosity bordering on the insane' (*FW* 173.34), is the spirit of the writing and the immediacy of the text. Wittgenstein wonders about the legitimacy of calling a secondary, corrective gesture – and I think the construction of these wobbly tables of contents is just that – an interpretation: 'Der Zerstreute, der auf den Befehl "Rechts um!" sich nach links dreht, und nun, an die Stirn greifend, sagt "Ach so – rechts um" und rechts um macht. – Was ist ihm eingefallen? Eine Deutung?' (139: 'The absent-minded man who at the order "Right turn!" turns left, and then, clutching his forehead, says "Oh! right turn" and does a right turn. – What has struck him? An interpretation?').

The third of my simple-minded questions is a favourite of the *Wake* itself: 'Are we speachin d'anglas landadge or are you sprakin sea Djoytsch?' (*FW* 485.12–13); the pronoun change here is often overlooked. *Finnegans Wake* and its critics – the scaffolding is only the most egregious case – do not speak the same language, and genuine attempts to reconcile the two tongues, rather than subordinate that of the text, are rare. Tindall, for one, refuses to acknowledge this difference: 'whatever the Breton and Telugu, words from such languages are rarely essential; for the *Wake* is "basically English" ([*FW*] 116.26) and Webster's dictionary, preferably the second edition, is our handiest guide' (Tindall 20). Checking the quotation, one finds a rather more conditional phrase, 'however basically English,' just as a few lines later one finds 'however basically Volapucky' (*FW* 116.31), which Tindall neglects, along with all other similar jibes about the text's 'most unenglish' (*FW* 160.22) nature. If *Finnegans Wake* is English, those who adopt this view tend to reason, then it is an English in need of repair. Margot Norris offers a representation of the comprehension-by-compensation dynamic when she claims that ungrammatical *Wake* sentences 'still communicate because the reader unconsciously recognizes the slot and knows the correct filler' (*The Decentered Universe* 127–8). Wolfgang Iser writes that the reader's decision as to how to fill 'gaps left by the text ... implicitly acknowledges the inexhaustibility of the text; at the same time it is this very inexhaustibilty that forces him to make his decision' (280).[4] But 'inexhaustibility' here can also connote the countless distortions that can be rendered, and Iser chooses not to see the violence inherent in the etymology of 'decision.'

A more cautious method of 'coming to terms' with the *Wake* is the three-step program advocated by keymasters Campbell and Robinson: 'The task of opening the way into any passage [of *Finnegans Wake*] ... divides itself into three stages: (1) *discovering* the key word or words, (2) *defining* one or more of them, so that the drift of Joyce's thought becomes evident, (3) *brooding* awhile over the paragraph, to let the associations running out from the key centers gradually animate the rest of the passage' (359–60). The assumption that the *Wake* is somehow closed I will address in a moment (and the question of the text's animation I will save for another chapter). The verbs here, as well as that nagging 'key' metaphor Levin wisely recommends dropping, seem (not least for their italicization) forced and inappropriate, particularly '*brooding*,' which is naturally reminiscent of that line of Yeats that haunts Stephen: '*And no more turn aside and brood*' (*U* 9). Just how much distinction there is to be made between the discovering-defining-brooding method and that of comprehension-by-compensation is difficult to ascertain.

I find another linguistic strategy in Bishop's *Joyce's Book of the Dark*. Examining the possibilities of the phrase 'this is nat language at any sinse of the world' (*FW* 83.12), Bishop postulates that if 'nat language' suggests 'not language,' 'the phrase indicates that the language of *Finnegans Wake* will work heavily by oppositional negation. Unlike English, that is, which conveys meaning in its ideal form by indicating the presence of corresponding ideas and things, this 'not language' operates largely by indicating their "Real Absence." The "lexical" parallels into which the "outlex" [*FW* 169.03] of Wakese can be translated, accordingly, indicate largely what the *Wake* is *not* about' (51; italics in original). That the *Wake* is about what it is not about complements well my earlier argument that the *Wake* is, or at least attempts to be, more than it is. These recognitions make the *Wake* no less intimidating, for every interpretive step forward into the text (as it were) is fraught with tripwires which detonate contradictions elsewhere in the book. Readers are free to associate, but Joyce offers no endorsement to these associations. Eco wonders:

> Can we speak of unlimited semiosis when we recognize the same technique [the exegesis employed in Hermeticism] implemented by contemporary readers who wander through texts in order to find in them secret puns, unheard-of etymologies, unconscious links, dances of 'Slipping Beauties,' ambiguous images that the clever reader can guess through the transparencies of the verbal texture even when no public agreement could support such an adventurous misreading? There is a fundamental principle in Peirce's semiotics: 'A sign is something by knowing which we know something more' ... On the contrary, the norm of Hermetic semiosis seems to be: 'A sign is something by knowing which we know something *else.*' (*Limits of Interpretation* 28; italics in original)[5]

Eco's pun, 'Slipping Beauties,' constitutes a sidelong look at the exegetical fervour brought to passages of *Finnegans Wake* such as this one: 'For it was in the back of their mind's ear, temptive lissomer, how they would be spreading in quadriliberal their azurespotted fine attractable nets, their nansen nets, from Matt Senior to the thurrible mystagogue after him and from thence to the neighbour and that way to the puisny donkeyman and his crucifier's cauda. And in their minds years backslibris, so it was, slipping beauty, how they would be meshing that way' (*FW* 477.18–24). Campbell and Robinson boil this down to: 'For it was in the back of their minds how they would be spreading their nets to mesh his issuing fish breath' (290). Tindall presents it as the 'first attempt on Yawn, that "slipping beauty"' (258). These glosses are certainly 'something *else,*' at least in the sense that they are English replacements for the troublesome texts. As liberating as it sounds, 'unlimited semiosis' can lead to derangement, that of the studied text when paraphrased (roughly, the narrative of interpretation refuses to work with the signs provided), and even that of the mind, which is the cautionary tale of Denis Breen (the imposed narrative demands signs that suit its purpose, rather than the other way around).

In looking over Eco's semantic/semiotic division of reading, Rabaté suggests that a 'pragmatic approach would be closer to Joyce's insight into an *auctoritas operis* which can be developed for its own sake. It would thus recommend not a semiotic interpretation, but only a semantic interpretation, developed when we learn to play with the text. The reading process becomes indeed a learning process, with its own specific pedagogy, in the way one learns (first making all sorts of errors) to play games such as bridge, scrabble or interactive video-programs' ('Pound, Joyce and Eco' 495). I would go further than Rabaté on the point of semiotic abstention and argue three seemingly small but very important points of difference by way of conclusion to this chapter and introduction to the two that follow.

My first contention is that *Finnegans Wake* defies Eco's well-known, convenient designations of 'open' and 'closed' work(s). The *Wake*'s hybridist nature extends beyond genre, even beyond portmanteau words, indeed to the most atomic level of text (letters, punctuation, sigla of all forms). Because what is 'clearobscure' (*FW* 247.34) is neither 'clear' nor 'obscure,' and yet is both, the unhalting dialectic, recognized in chapter 6, between 'error' and 'irony' within a text is also in operation between what a text may signify and what it may not. In the *Symposium*, Alcibiades 'compares Socrates with the Sileni, those carved figurines with satyrlike and grotesque images on the exterior, but pure gold inside' (Behler 80). While reading *Finnegans Wake* might be the learning experience Rabaté suggests – it may even be an 'indoctrination' like that from which McHugh recoils – its quasi-Socratic pose offers no guarantee of 'pure gold,' or anything at all, inside.

I also have some scepticism about the semantic-semiotic reader distinction proposed by Eco, because I am unwilling to concede that semantics – or even, if you like, the very look of the print on the page as it is glimpsed by an illiterate – possesses no opportunities for semiotic or critical consideration. *Finnegans Wake*'s project to 'keen again and begin again to make soundsense and sensesound kin again' (*FW* 121.14–16) challenges this distinction, leaves us with what might be called the Dogberry Question after the delicious double-malapropism made by the constable in *Much Ado About Nothing*: 'our watch, sir, have indeed comprehended two aspicious persons' (3.5.45–6). At this point in the play, the defaming (suspicious or auspicious) villains have been caught (apprehended or comprehended) by the clowning agents of justice. Yet can it be said that Dogberry is wrong to suggest that the villains have been 'comprehended'? How tangible is the separation between the process of seizing upon something, be it rogue or text, and recognizing it, as 'aspicious' or otherwise? Guy Davenport writes: '[y]ou do not read *Ulysses*; you watch the words' (*Geography* 287), to which can be added, in the case of *Finnegans Wake*, one watches the moving letters, watching like Dogberry's lieutenants, who are told: 'to be a well- / favoured man is the gift of fortune, but to write / and read comes by nature' (3.3.14–16). I will have more to say about this 'nature' in the next chapter.

Third, and finally, is my rejection of the mollifying 'first' in the phrase 'first making all sorts of errors,' which itself is offered in parentheses, like a shameful thing. To read *Finnegans Wake* is to make mistakes, and to enjoy the *Wake* is to cherish what the mistakes reveal. Interpreting by 'trial and error' is, in the words of Fritz Senn, 'part of our survival strategies' (*Joyce's Dislocutions* 42), or, as Benstock calls it, 'the prescribed if precarious method' (42). It seems to me that the reader of *Finnegans Wake* faces a paradoxical situation very similar to that of someone attending a performance of Tom Johnson's notorious 1975 composition, 'Failing: A Very Difficult Piece for Solo String Bass.'[6] Ostensibly a soloist's piece, 'Failing' is made up of a musical score and a written text for accompanying monologue. The text, which is to be read at a steady, calm pace, as often as not at odds with the varying string bass rhythms, itself is an explanation of and meditation on the composition's strategies. While the player tries to juggle the contrary demands of playing and speaking (interacting with two different kinds of text), the matter of the speech concerns the anxieties the performer inevitably has in attempting what is by its own admission 'Very Difficult' and very likely impossible. What makes 'Failing' most interesting is its inclusion of the audience in the self-consciousness of the performance. When the player contends, 'I have practised the piece quite a bit, and that's a fact, as well as just simply a line in the text,' the audience may wonder about the separation between the volition of the musician and that of the text before him or her. Again, the Dogberry Question in a different scenario: where does the interpretation (the player's *and* the audience's) begin? The player continues: 'I should point out, however, that I am not obliged to fail. After all, the audience cannot see the score I am reading and no one would ever be the wiser if I were to simply leave out whatever passages I am unable to play. People might be very impressed by my playing and think that I had succeeded in playing a piece which the composer had thought could not be played successfully by any bass player.' The audience of 'Failing' is made to realize that the pressures felt by the player emanate, at least in part, from themselves, and the performance of 'Failing' projects it back to them, so that they cannot think of themselves as outside or external to the performance and its concerns. Readers of *Finnegans Wake* likewise find themselves contained within the process of interpreting the text. The idea that the performance of interpreting *Finnegans Wake* culminates in 'understanding' is equivalent to the idea of 'successfully' playing Johnson's 'Failing.'

Robert Sage's thesis that in the *Wake* Joyce 'brings to fruition what was foreshadowed in *Ulysses*; the possibility of a complete symbiosis of reader and writer' (143), can be best appreciated if the performance of reading the *Wake* is not separated from its accompanying anxieties.

CHAPTER EIGHT

Fickling Intentions (II)

> You remember fairly accurately all my errors, boasts, mistakes.
>
> (*U* 622)

'Misreadings,' by virtue of being interpretations, can cast as potent a charm – or as sinister a curse – as any other (more readily arguable or academically or authorially sanctioned) interpretations upon the idea of a text. Encountering now the bitter epithet 'phony' in *The Catcher in the Rye*, for instance, may well produce an irresistible echo of a 1980 gunshot in wintry New York, even though Salinger's novel can nevertheless be understood as anything but a programmatic directive for celebrity assassination. Only very recently has the music of Wagner been given its first public performance in the state of Israel, where it has been hitherto silently thought of with only grim associations. Legacies of interpretation have varying lifespans (those of scripture are collectively the most obvious example of longevity and resilience as well as of change and diversity), many of which (like any other lifespan) are not lacking for suffering and misfortune, but the reading history of a text enshrouds it always. The ghosts of erroneous interpretations – and all interpretations, I propose to argue in this chapter, can be thought of as errors – mean that reading is a haunted act.

Heidegger most austerely stated that the philosopher's duty is to listen to the silence of existence. If there is any tenable analogy to be had here, to or for what does the literary critic listen, and how? Can we hear the ghosts of past readings and readers in our own experiences of interpretation? The metaphor in Yeats's refrain, '*Like a long-legged fly upon the stream / His mind moves upon silence*' (381–2), provides an entry point for this search for a philosophic-poetic interface. The Gripes of *Finnegans Wake*, 'having the juice of his times' is likewise 'parched on a limb' by a stream (*FW* 153.9–11), as much a fly as a Narcissus, readily drinking away and yet thirsty. (I will bring yet another insect into focus shortly.) Both Yeats and Joyce combine, or at least refuse to distinguish between, the sensory acuity of the artist and that of a work's reader or audience, or even between impression and expression. The latter phenomena – my reasons for refraining from calling them acts will become apparent in due course – *together* constitute interpretation, but as *Finnegans Wake* reminds us, we readers of literature have 'Fickleyes and Futilears' (*FW* 176.13): experiential prejudices, conditioned responses, and habituated if not dulled sensory receptors – more than perhaps we care to admit or know.

Linda Hutcheon, in her study of irony noted in chapter 6, cautions that 'intentional/non-intentional may be a false distinction: all irony happens intentionally, whether the attribution be made by the encoder or the decoder. Interpretation is, in a sense, an intentional act on the part of the interpreter' (118). There is here a simple but vital – in the sense that it inspires constructive debate – point dressed in some questionable terminology; the hesitation in that otiose phrase, 'in a sense,' allows for a total negation of the proposition, and the distinction between 'the encoder and the decoder' is one that chooses to overlook the fact that, in cryptological terms, an interpretation of a text is not a translation of the text's 'code' into a non-codified language (this is most starkly demonstrated in bad academic writing). Readers have intentions, but what does this statement *mean*?

Let us begin with a crude model of how 'fickling intentions' operate within reading and move gradually towards complexity and sophistication. Jerry R. Hobbs constructs the following 'formula' of interpretation:

$F(K, T) = I.$

An interpretation procedure F takes a knowledge base or belief system K and a text T, and produces an interpretation I ... In general, there should be little dispute about T ...

Ultimately, in text interpretation as in every scientific or critical enterprise, we must bottom out conventionally agreed-upon 'evidence.' For text interpretation, this first involves a decision or an agreement *that* some physical object exists or *that* some physical phenomenon has occurred. (20–1; inexplicable italics in original)

My own interests in potential interplay between literary and cognitive studies, which fuel the next chapter, do not prevent me from squirming at the rigidity of this reductive, mechanistic approach to both literature and the human mind. The phrase 'simple-minded' has special application here. As it turns out, though, in his next two sentences Hobbs effectively makes my counter-argument for me: 'I doubt that any literary critic, as a critic, could seriously maintain that copies of *Ulysses* do not exist as physical objects, regardless of what one may take them to be. If we cannot accept the reality of trees, chairs, and books, it is hard to see why we should care about the feelings of Stephen Daedelus toward Leopold Bloom' (21). That pedantic qualifier, 'as a critic,' is lightly stepped over by any reader who notes how 'Dedalus' is spelled,[1] and the logic of the last sentence is highly questionable. Hobbs has an inconsistent notion of the importance of a text's physicality. On the one hand, it is for him an essential, the *donnée*, of interpretation – a view that would invalidate certain major tenets of religion, philosophy, and psychology – but on the other hand, the informational qualities of physical text, or indeed any general awareness of medial effects, are absent in considerations of variables K and T (about which there is 'little dispute').

E.D. Hirsch offers similar banalities, though they usually appear more literate. In his argument for 'Objective Interpretation,' Hirsch declares that '[i]f criticism is to be objective in any significant sense, it must be founded on a self-critical construction of textual meaning, which is to say, on objective criticism' (27), a ludicrous sentence that can quickly be pared down: 'If criticism is to be objective ... it must be founded ... on objective criticism.' While Hirsch is against 'the view that a text is a "piece of language"' – and rightly, since a literary work has further dimensions than a linguistic act, and in any case both 'text' and 'piece' are inappropriate, even clumsy, terms with which to discuss this dimension – he argues instead for 'the notion that a text represents the determinate verbal meaning of an author' (39). Since 'verbal meaning,' in his estimation, 'must conform to public linguistic norms (these are highly tolerant, of course), no mere sequence of words can represent an actual verbal meaning with reference to public norms alone. Referred to these alone, the text's meaning remains indeterminate ... Just as language constitutes and colors subjectivity, so does subjectivity color language. The author's or speaker's subjective act is formally necessary to verbal meaning, and any theory which tries to dispense with the author as specifier of meaning by asserting that textual meaning is purely objectively determined finds itself chasing will-o'-the-wisps' ('Objective Interpretation' 39). Note that the last sentence actually offers *two* distinct propositions, not one, as its rhetorical shape tries to suggest. Passing over the 'specifier of meaning' whose toe Barthes most suavely tagged is, of course, not causally connected with the assertion of 'purely objectively determined' meaning. Hirsch insists that 'interpretation is the construction of *another's* meaning' – this he calls a 'half-forgotten truism' ('Objective Interpretation' 54; italics in original) – but this apparently rigid definition is not altogether transparent, since the degree of activity in this feat of 'construction' is unspecified. Is the reader a passive vessel (not necessarily empty, since to be so would validate McHugh's claim for an 'innocent' reading, which I have already rejected) that is affected or altered when it receives and contains the reactive chemical of the text ('Joyce does this to me'), or else the active agent, arguably imperial in gesture, circumscribing in rough strokes that ostensibly cherished but contained virtue of the classic ('I understand why Joyce does this')?

Roughly, the question is whether the reader can somehow negate him- or herself as he or she reads. Although it may be tempting in a climate of pronounced positionality discourse to deride it, this problem has never truly left scrutiny. William Elford Rogers proposes the possibility of a hermeneutic state of zen wherein 'the interpreter purges from consciousness purely private feelings and awareness of the separate "I" of the "I think." That is, the interpreter aims at a pure consciousness of the representative function of the signs woven together to make the text' (135). As the subtitle of his 1994 book explains, *Interpreting Interpretation: Textual Hermeneutics as an Ascetic Discipline* is a protraction of a formidable (though I think flawed) simile. Ascetic discipline, writes Rogers, is 'like being totally absorbed in a conversation, to the point that words come without seeking, and one is not aware of any separation between the words and the meaning. Or it is like fluent reading, where there is no sense of having to decode the marks on the page and having no sense of having to extrapolate from what is written to some absent authorial intention. One reads, and the meaning comes' (132). If all this sounds uncomfortably like Valley Girl speech ('like being totally absorbed ... like fluent reading ... having no sense of having'), the sensation may be indicative of the scarcely concealed wistfulness of the enterprise. Textual history rides with individualism's development, though Marshall McLuhan would point out that the pluralities of text make the experience of them a virtual schizophrenia, recognized even within literature from print-mad Quixote to the polynominal mutterings of the *Wake*. Rogers is circling the Dogberry Question in his own medeoturanian manner, however, and he implies that reading can and does occur involuntarily. In the last sentence I employed a word randomly selected from *Finnegans Wake* (289.20), a word that has no apparent meaning in the sentence, and yet it can be read, given a pronunciation and a context or framework of possible (and/or impossible) meanings. This reflex is the literal instinct. In his animal analogies that I quoted in the last chapter, Wittgenstein questions the division between a rational or even conscious intention and an impulsive or unconsciously compulsive act. To alter the last sentence from Rogers I quoted above: one reads, and the meaning is expected to come. In this way the Gutenberg Galaxy has indelibly conditioned us.

Because textual signs are so desperately seized upon by the literate even when the signs themselves only appear to be recognizable or otherwise translatable (Chomsky's 'Colorless green ideas sleep furiously' [15] is a demonstration of syntax's separation from sense), Hirsch's distillation of Saussure seems shortsighted. In the dictionary that Hirsch privileges as a controlling force over poetry, '[t]he letters in boldface at the head of [a] definition represent the word as *langue*, with all its rich meaning possibilities. But under one of the subheadings, in an illustrative sentence, those same letters represent the word as *parole*, as a particular, selective actualization from *langue* ... Of course, many sentences, especially those found in poetry, actualize far more possibilities than illustrative sentences in a dictionary. Any pun, for example, realizes simultaneously at least two divergent meaning possibilities. But the pun is nevertheless an actualization from *langue* and not a mere system of meaning possibilities' ('Objective Interpretation' 45).[2] The last sentence demonstrates that Hirsch never looked deeply into *Finnegans Wake*, where he would have found his position chastened like that of 'a lexical student, parole, and corrected with the blackboard (trying to copy the stage Englesmen he broughts their house down on, shouting: Bravure, surr Chorles! Letter purfect! Culossal, Loose Wallor! Spache!)' (*FW* 180.36–181.03). It also suggests that, despite his lip service given to the idea of its rejection, Hirsch cannot entirely give up the notion of text as principally a linguistic act.

The question of how pervasive or common is this 'textual conditioning' needs to be studied against the matter of interpretive diversity. Wolfgang Iser concludes his book *The Implied Reader* by qualifying individuality as it operates in textual interpretation: 'The impressions that arise as a result of [the process of reading] will vary from individual to individual, but only within the limits imposed by the written as opposed to the unwritten text. In the same way, two people gazing at the night sky may both be looking at the same collection of stars, but one will see the image of a plough, and the other will make out a dipper. The "stars" in a literary text are fixed; the lines that join them are variable' (282). No analogy could be more appropriate and yet so misconstrued. The immobility of the stars is an illusion; all perception is bound up in relative motion, and I argue that this includes the act of reading. Bloom's pet word, 'parallax,' echoes again here; but if I can leave Joyce for a moment, there is another literary case that offers a stronger rebuff of this metaphor as well as a dramatization of another important problem of interpretation.

When guilt-ridden Arthur Dimmesdale in *The Scarlet Letter* witnesses within the spectacle of a meteor shower 'the appearance of an immense letter, – the letter A, – marked out in lines of dull red light' (175), Hawthorne's narrator slyly questions Dimmesdale's anguished state of mind by casting doubt on 'the faith of some lonely eyewitness, who beheld the wonder through the colored, magnifying, and distorting medium of his imagination, and shaped it more distinctly in his afterthought' (174). The 'distorting medium of [the] imagination' is understood, then, to be the handicap of the individual. Quality time with one of Stanley Fish's interpretive communities would straighten wretched Dimmesdale out – think of such an event as a healing session of Overinterpreters Anonymous – but Hawthorne himself remains ambiguous as to whether or not interpretive consensus constitutes simply a stabilizing force or an oppressive one, just as he does not openly recognize the celestial 'A' is anything other than a perception.

Hawthorne and Joyce compare the letter with the identity of those interpretive forces or presences to deal with their immediacy. *Finnegans Wake*, which does not follow the customary 'gentle' or 'dear reader' form of address but rather defers to 'drear writer' (FW 476.21), 'ye aucthor' (FW 148.17), 'gentlewriter' (FW 63.10), and so on, thrusts the pen of judgment into the reader's adulterating hands, to write new 'haunting hesteries' (FW 319.06–7). The trial is the reader's as much as it is the text's. As readers, we are all the same in that we are *not* by dint of individual experience, just as our interpretations – not including but constituted by our errors – are uniformly different.

Although I am proposing that errors make for the diversity of textual interpretation, I am not arguing that all interpretations are equally 'valid'; for to do so would be equivalent to saying that no interpretation is valid. The diversity of Joyce (author configurations and texts) and readers ensures a diversity of interpretations, however, and, as I have suggested that the 'meaning' of a text is in the interrelations of these interpretations, no single interpretation is in this sense 'correct.' Error's non-equivalency makes each of us a Dimmesdale, haunted but individual.

What about reading approaches: do they at least have limits? In *Interpretation and Overinterpretation*, Umberto Eco refers to 'jumping here and there through the text' unflatteringly as 'grasshopper-criticism' (72), but this sobriquet would hardly disturb a *Wake* reader, who knows that the excesses of that particular insect provide greater pleasure than the 'hardworking straightwalking stoutstamping securely-sealing' (*FW* 603.10–11) linear toil of an ant. One need only spend a little time with well-known books such as *Tristram Shandy*, *Hopscotch*, or any of Gertrude Stein's writings to notice that 'jumping around' is what one does in such textual obstacle courses, though these examples are not merely showy extremes that prove the ant's rule of beginning-to-end reading but are, rather, participants in literature's debate with itself on such issues. Joyce, one of the most active of these participants, was wondering about anti-linear hermeneutics years before he even conceived of the *Wake*, which advises that 'the words which follow may be taken in any order desired' (*FW* 121.12–13). As a schoolboy Stephen Dedalus turns over a number of linguistic questions, and the ones that seem most facile are actually both the most complicated and the focal sites of Joyce's disruptive poetics. Stephen looks at words written about himself by another –

Stephen Dedalus is my name,
Ireland is my nation.
Clongowes is my dwellingplace
And heaven my expectation

– though '[h]e read the verses backwards but then they were not poetry' (*P* 16). One wonders precisely what 'backwards' route Stephen takes (by line, word, or letter?), but the word 'poetry' is jarring, disciplinary, like the 'smack' of the pandybat of Father Dolan (*P* 49–50). The young man's later preference for Byron, 'a poet for uneducated people' (*P* 81) like those people who read verses backwards and thoughtfully jump 'here and there through the text,' over the 'rhymester' Tennyson (*P* 80) illustrates Stephen's first grasp of the principle of definition by redefinition.

Donald Theall's notice of writing as an act that 'involves a method of dialectical doubt ... It is a "hophazard" [cf. *FW* 615.07–8] process, a process of chance' (208) could also be applied to reading, with the understanding that the process of chance is in operation at every moment of reading, even in the selection of text. By 'selection of text' I mean what strikes one's eye – think of Stephen's reaction to the unanticipated graffito '*Foetus*' [*P* 89]) – and the ambiguity of the phrase permits the consideration of where the agency of selection resides. *Finnegans Wake* refers to 'jaywalking eyes' (*FW* 121.17), both the fickle gaze of the hapless reader and the incautious letters on the page themselves, fugitive 'J's and 'I's scrambling about in search of other letters and readers.[3] When the same text asks, 'Why such an order number in preference to any other number? Why any number in any order at all?' (*FW* 447.25–7), it seems to invite the reader to challenge sequence as a guidance system.[4] The order of poems in *Chamber Music*, which I discussed earlier, or in the later, neglected *Pomes Penyeach* is not as readily explained as a metanarrative structure as the maturation schema applied to *Dubliners*, so it is rarely discussed. It could be argued, however, that the lack or apparent looseness of sequential structure itself is a refutation of such impositions. The last stanza of a poem called 'Flood' runs:

> Uplift and sway, O golden vine,
> Your clustered fruits to love's full flood,
> Lambent and vast and ruthless as is thine
> Incertitude! (*Pomes Penyeach* 653)

The incertitude of which the reader, besides the narratee, is accused is indeed, in Joyce's greatest works 'vast and ruthless' – not least because the works themselves are – but also (that warm, wonderful word) '[l]ambent': gently illuminating and also with the light touch of gracehoper's wit.

Let me jump back in my own text to Eco, who goes on in *Interpretation and Overinterpretation* to muse on fallacious associations and contexts (not his terms, though I think they fit) and comes up with this example: 'Jeanne d'Arc was born in Domrémy; this word suggests the first three musical notes (do, re, mi). Molly Bloom was in love with a tenor, Blazes Boylan; blaze can evoke the stake of Jeanne, but the hypothesis that Molly Bloom is an allegory of Jeanne d'Arc does not help to find something interesting in *Ulysses* (even though one day or another there will be a Joycean critic eager to try even this key)' (77). If 'try' may be taken to mean 'to examine judiciously,' I volunteer to do so, because Eco's creative example of a misreading, or what he would term 'use' rather than 'interpretation,' is not a fair one, flawed as it is by the fact that one of its connective assumptions itself is a misreading. Whatever else he may be, Blazes Boylan is not a singer (Eco perhaps confuses him with Simon Dedalus, who sings in the 'Sirens' episode, or supposes that Boylan's arrangements for Molly's concert suggest a duet). Neither is it entirely evident that Molly is 'in love' with him, though they are enjoying a sexual affair. It is true that Molly likes a good tenor (they do 'get women by the score' [*U* 353]), and Bloom jealously thinks of Bartell d'Arcy '[s]eeing her home after practice. Conceited fellow with his waxedup moustache' (*U* 196–7). D'Arcy, who is included in the 'series' of Molly's lovers indexed in 'Ithaca' (*U* 863), has a name that links him to Jeanne from Domrémy. Whether this connection is 'interesting' in regard to *Ulysses* is naturally contestable, but what I have offered – an echo of Eco with a difference – is not a fully developed argument but a few connective points preliminary to an expressible interpretation. In Eco's view, such an interpretation ought to be 'a conjecture about the *intentio operis*' that can be proved by 'check[ing] it upon the text as a coherent whole' (*Interpretation and Overinterpretation* 65). I, on the other hand, am in agreement with Richard Rorty in finding the metaphor of a text's '*internal* coherence' untenable (97; italics in original) and would even cast doubt on discussion of a text's *external* coherence (the shroud of meaning I gestured to earlier) as anything other than an abstraction.

Finding contexts thus seems to me less a threat than losing, limiting, or (as Eco does to poor Joan of Arc) martyring them. As Derek Attridge argues, constant challenges to interpretive contexts are vital because they 'help to free interpretations for fresh scrutiny of the text and new thinking about its implications. The same cause is aided by playing down the significance of authorial comment – either by a theoretical argument against intentionalism or by a sceptical examination of reported comments in their historical context. It is likely that the new interpretations thus made possible will be less geared to a project of assimilating and regulating what is difficult and unorthodox. But what we cannot claim is that we are replacing *error* with *truth* (*Joyce Effects*

150; italics in original). What we need is a circulating and varying range of contexts, but Rabaté notes how a text of such complexity as the *Wake* offers no alternative comfort to that of the seductive desire to make the very claim Attridge cautions against: 'Precisely because a continuous insomnia generates a symptomatology, the "ideal genetic reader" will be offered a choice between varieties of error, and typologies of pathological readings. One could distinguish between obsessional structures (the work will have to be studied rigorously, with a minimum of help from outside sources), paranoiac delusions (which often yield the sense of having found the ultimate key or cracked the code) or hysterical projections (the text will become a master who will have to be seduced at any cost, often at the risk of losing any regard for decency or common-sense)' ('Pound, Joyce and Eco' 494–5). All of these critical concerns come to this: it is very hard to have a healthy relationship with *Finnegans Wake*. Yet while Eco cuts down an unfinished straw man and Attridge and Rabaté theorize about general tendencies, none of these commentaries gives any concrete examples of any *Wake* interpretation, faulty or otherwise. Before offering a speculation on the meaning of this evasion, let me first solidify the point that the technique *is* an evasion by consulting a pair of examples, one an unscholarly interpretation and one by an estimable Joycean.

Noel Riley Fitch reports how in 1954 a young American named George Johnson 'sent letters and telegrams announcing that he had "solved the riddle" of *Finnegans Wake*: When would World War III break out? On page 517, he discovered, Joyce declares, "Tick up on time. Howday you doom? ... The uneven day of the unleventh month of the unevented year. At mart in mass." Atomic war, [Johnson] concluded, was going to start on 11 November (Martinmas). In forty letters to Sylvia, his family, and his friends, he demanded that they evacuate the cities' (394). 'Bilking' in the *Wake* is not only the gesture of the writer, but of the reader: 'So read we in must book. It tells. He prophets most who bilks the best' (*FW* 304.31–305.02). When the twelfth of November dawned in 1954, George Johnson must have felt thoroughly bilked, for he propheted nothing. His reading seems extraordinary, not least of all because his investment in it risks more than an opportunity for university tenure, but in the vast realm of possible misreadings, which I will refer to after Joyce as 'Errorland' (*FW* 62.25), this is but a twinkling, a bizarre but comparatively unsophisticated and unelaborate mis-taking of the text.

My second example is also concerned with numbers. Margot Norris finds in a passage from the 'Night Lesson' of the *Wake* 'an obvious error': '"Thence must any whatyoulike in the power of empthood be either greater THAN or less THAN the unitate we have in one ...' [*FW* 298.11–14] suggests that any number ("whatyoulike") raised to the power of zero ("power of empthood," 2^0, for example) must be greater or less than one. A number raised to the power of 0 is, of course, equal to 1. Since the entire paragraph comprises a theorem, we may assume that an error in one part also reverses other elements in the theorem' (*Decentered Universe* 46).[5] Putting aside the specious logic of the last sentence, it is obvious that Norris's reading here is selective. Translating 'empthood' as 'zero' is no less specious than finding the eleventh of November in '[t]he uneven day of the unlevnth month.' Anthony Burgess notes that 'Joyce often spoke of [*Finnegans Wake*] as mathematical, and one thing in it that the vast chaotic dreaming mind never impairs is number' (*Here Comes Everybody* 270), but the truth of this statement clearly depends on your definition of 'impairs.'

Norris's explication of the 'obvious error' omits mention of the text's own conclusion to these exponential thoughts: 'Which is unpassible. Quarrellary. The logos of somewome to that base anything, when most characteristically mantissa minus, comes to nullum in the endth ... Scholium, there are trist sigheds to everysing but ichs on the freed brings euchs to the feared. Qued? Mother of us all! I don't know is it your spictre or my omination but I'm glad you dimentioned it! (*FW* 298.18–299.06). Adopting Norris's approach one could suggest that here the mathematical principle,

$$\text{for any } n \parallel 0 < n \in \mathbb{R}, \log_n(1) = 0,$$

is correctly represented, given that 'somewome' is glossed as 'one,' 'nullum' as 'zero,' and so on. But all of this is literally much ado about nothing. Nothing is proved (not 'QED' but 'Qued?'), and each obscure line of text builds upon the 'impassible' truth of the last: impossible; a failure; but permanent (an arrogant appropriation of Matthew 24:35). Robert Adams quips: 'arithmetic was never [Joyce's] long suit' and mathematical errors such as these 'are not simply parodic; there is a kind of bumbling logic behind the whole thing,' but he doubts that such an imponderable scheme justifies 'errors of primitive simplicity' (183). The *Wake* retorts that it openly favours variation and mutation: 'if you are not literally cooefficient, how minney combinaisies and permutandies can be played' (*FW* 284.11–13).

Norris finds fault with the text for its failure to conform to her arithmetic; Johnson finds fault with the apocalypse, which failed to happen on cue from the text's stage direction. The reason such examples are seldom pointed to in studies of the *Wake* is that no critic wishes to condemn categorically any interpretation of this text, to say that a reading is 'wrong.' The *Wake* never ceases to exploit these 'hesitensies' (*FW* 187.30): 'I need not anthrapologise for any obintentional (I must here correct all that school of neoitalian or paleopraisien schola of tinkers and spanglers who say I'm wrong parecqueue out of revolsican from romanitis I want to be) downtrodding on my foes ... (I am purposely refraining from expounding the obvious fallacy as to the specific gravitates of the two deglutables implied nor to the lapses lequou asousiated with the royal gorge through students of mixed hydrostatics and pneumodipsics will after some difficulties grapple away with my meinungs)' (*FW* 151.07–11; 26–31). The 'students of mixed hydrostatics and pneumodipsics' who come to breathe or drink in the *Wake* will be overloaded, gorged on possibilities. Discrimination is not what it used to be.

Eco contrasts 'possible worlds that sound *nonverisimilar* and scarcely credible from the point of view of our actual experience' in which the reader acts 'as a nearsighted observer able to isolate big shapes but [is] unable to analyze their background' with '*impossible possible worlds*, that is, worlds that the Model Reader is led to conceive of just to understand that it is impossible to do so' (*The Limits of Interpretation* 76; italics in original). In the case of 'possible worlds' the text is understood a priori to be erroneous; thus, it is granted exceptions (talking animals, flying carpets, transversal mirrors, violations of mathematical laws). In the case of impossible possible worlds, the text instructs in the course of its being read the erroneous nature of its own matter. If the latter case seems to apply to the *Wake*, what kind of 'world' is *Ulysses*?

Joyce's verisimilitude is celebrated even by those readers who (usually when they arrive at the 'Circe' episode) cannot finish *Ulysses*. Although he undergoes some startling transformations in 'Circe' – and these can be rationalized as drink-fuelled hallucinations – Bloom is never in two places at the same time, as he might be in a fiction by Borges or Robbe-Grillet, and this unity is often understood to be, in the particular 'local, nonhomogeneous small world' of Dublin on 16 June 1904, irresistible. This unity may be the most important facet of mimesis within the novel, if mimesis is understood to be the inferable parallelism of cosmogonic metaphors. Yet its violation is significantly hinted at in the unusual and comic dispute over the whereabouts of Paddy Dignam in the 'Cyclops' episode:

>   –How's Willy Murray those times, Alf?
>   –I don't know, says Alf. I saw him just now in Capel Street with Paddy Dignam. Only I was running after that ...
>   –You what? says Joe, throwing down the letters. With who?
>   –With Dignam, says Alf.
>   –Is it Paddy? says Joe.
>   –Yes, says Alf. Why?
>   –Don't you know he's dead? says Joe.
>   –Paddy Dignam dead? says Alf.
>   –Ay, says Joe.
>   –Sure I'm after seeing him not five minutes ago, says Alf, as plain as a pikestaff.
>   –Who's dead? says Bob Doran.
>   –You saw his ghost then, says Joe, God between us and harm.
>   –What? says Alf. Good Christ, only five ... What? ... and Willie Murray with him, the two of them there near whatdoyoucallhim's ... What? Dignam dead?
>   –What about Dignam? says Bob Doran. Who's talking about ...?
>   –Dead! says Alf. He is no more dead than you are.
>   –Maybe so, says Joe. They took the liberty of burying him this morning anyhow. (U 388)

The reader cannot positively confirm or deny that Alf Bergan, whom Bloom guesses to be the author of the 'U.P.' postcard, saw Paddy Dignam in Capel Street, because the reader is not privy to such a scene. This is a remarkable lacuna in a novel that so exhaustively provides data and has its own kind of global positioning system by which characters' whereabouts can be confirmed. Alf's response that Dignam 'is no more dead than' Joe is a metafictional wink in that it is correct: Dignam the fictional construct is as alive or as dead as any other fictional construct.

Beckett gives the significant explanation that 'to Joyce reality was a paradigm, an illustration of a possibly unstatable rule' (qtd in Ellmann 551). The paradigmatic metaphor of 'reality' in *Ulysses* is by and large consistent with that of the world outside the reader's window, a metaphor governed by the natural sciences, but Joyce allows for a development of subversive submetaphors in his world, which submetaphors in

Finnegans Wake operate on a letter-by-letter basis of disintegration. Their presence and interaction trouble readers' assumptions of how mimesis operates. To trace one network of submetaphors – one rift or rupture in the clean mimetic surface – involves 'grasshopper-criticism' at its most energetic, and I will give one example.

In 'Scylla and Charybdis,' the chapter of dialectic and literature, Stephen produces an odd body count in his interpretation of *Hamlet*: 'Nine lives are taken off for his father's one, Our Father who art in purgatory' (*U* 239). Robert Adams calls this 'a curious, and, for all one can tell, a deliberate error' (129). Only eight (as if that were not enough) of the dramatis personae die in the course of the tragedy (in order of descent: Polonius, Rosencrantz and Guildenstern, Ophelia, Gertrude, Laertes, Claudius, and Hamlet). Stephen does not imply that Hamlet the king is the ninth, unless 'his father's one' refers to Shakespeare's father, the butcher – but this idea is sketchy. There is, however, an unusual echo of Stephen's afterthought, 'Our Father who art in purgatory,' in the 'Wandering Rocks' episode: 'Never see him again. Death, that is. Pa is dead. My father is dead ... Poor pa. That was Mr Dignam, my father. I hope he is in purgatory now because he went to confession to father Conroy on Saturday night' (*U* 324). Poor, befuddled Dignam the younger is the Hamlet Stephen affects to be, and '[n]ever see him again' is a prosaic variation of 'I shall not look upon his like again' (*Hamlet* 1.2.188), though as a pedant Stephen might be tempted to call it an error. *Finnegans Wake*'s reference to 'the Dane and his chapter of accidents' (*FW* 452.03) may be not only to *Hamlet* but to Joyce's own use (Eco's term for misinterpretation, you will remember) of Shakespeare's play, since it also recalls Bloom's summary of *Ulysses*: 'an unusually fatiguing day, a chapter of accidents' (*U* 630). Also, though 'grasshopper-criticism' itself can be unusually fatiguing, it has many accidental rewards.

Our fickling intentions as readers are in flux, rather than left as static prejudices, when reading Joyce, because we are never convinced that we are not misreading Joyce. *Finnegans Wake* calls for a sensory recalibration, an evolutionary acquisition of what Slingsby calls 'reciving cells' (90): a good portmanteau of receiving and reciting. In the light of the challenges posed by works such as the *Wake*, we need to rethink assumptions of 'order' as a principle of either aesthetics or cognition – these are the domains of the next chapter – and how concurrent shifts in textual production and poetics reshape our receptive abilities (would – or even *could* – a contemporary of Purcell detect anything but 'background noise' in Coltrane's saxophone?). Rows of radio telescopes across the North American continent lean forward to catch even a whisper of informational transmissions from beyond our planet. In this search, the recognition of 'intelligence' – the *telos* of these scopes[6] – appears as a trawling for predetermined signs of value. The search is predicated upon a metonymy between intelligence (typically anthropocentrically circumscribed) and order, usually in the form of patterns and logic-based designs. Yet whether sameness or difference is the ideal, or at any rate the more instructive, starting point for hermeneutics, is a question *Finnegans Wake* tantalisingly leaves open.

The 'other world' that Martha Clifford mentions (*U* 95) and that troubles Bloom is the world of error, the messages that seem scrambled and unreadable and are discarded as unintelligible and thus unintelligent. Our bias, as Blanchot points out, lies in our idolatry of 'truth':

> Par rapport au monde où la vérité a son assise et sa base toujours à partir de l'affirmation décisive comme d'un lieu où elle peut surgir, [l'art] représente originellement le pressentiment et le scandale de l'erreur absolue, de quelque chose de non vrai, mais où le 'non' n'a pas le caractère tranchant d'une limite, car il est plutôt l'indétermination pleine et sans fin avec laquelle le vrai ne peut frayer (Blanchot 326)

> (In the world, decisive affirmation dependably serves truth as a basis and foundation, as the place from which it can arise. By comparison, art originally represents the scandalous intimation of absolute error: the premonition of something not true but whose 'not' does not have the decisive character of a limit, for it is, rather, brimming and endless indeterminacy with which the true cannot communicate. [Smock 243])

Blanchot has, with startling concision, demonstrated the negative, wronging function of art. The *Wake* releases its 'not' like a thunderbolt – 'bababadalgharaghtakamminarronnkonnbronntonnerronntuonn-thunntrovarrhounawnskawntoohoohoordenenthurnuk!' (*FW* 3.15–17) – and we as readers find ourselves as much lightning rods as we may be telescopes. Its voltage offers a different information, knocks the 'e' out of 'receiving,' and waits for a response. The imperative underlying this phenomenon is suggested in this luminous passage in Proust:

> les vérités que l'intelligence saisit directement à claire-voie dans le monde de la pleine lumière ont quelque chose de moins profond, de moins

nécessaire que celles que la vie nous a malgré nous communiquées en une impression, matérielle parce qu'elle est entrée par nos sens, mais dont nous pouvons dégager l'esprit ... Les idées formées par l'intelligence pure n'ont qu'une vérité logique, une vérité possible, leur élection est arbitraire. Le livre aux caractères figurés, non tracés par nous, est notre seul livre. (Proust 3:878, 880; ellipsis added)

(the truths which the intellect apprehends directly in the world of full and unimpeded light have something less profound, less necessary than those which life communicates to us against our will in an impression which is material because it enters us through the senses yet has a spiritual meaning which it is possible for us to extract. ... The ideas formed by the pure intelligence have no more than a logical, a possible truth, they are arbitrarily chosen. The book whose hieroglyphs are patterns not traced by us is the only book that really belongs to us. [Moncrieff and Kilmartin 3:912, 914])

To what extent *Finnegans Wake* alone belongs to us – and we to it – I leave to the next chapter.

CHAPTER NINE

The allriddle of it

> The question within the question. To which does the question mark refer? If one question mark is lost, where does its meaning go? How is it possible for punctuation to have multiple or non-specific references?
>
> (Silliman 55)

> Ask yourself the answer, I'm not giving you a short question.
>
> (FW 515.19–20)

I

Let me come to the *'Punkt'* (*U* 261) by way of punctuation; as many questions as a given reader of any experience with *Ulysses* or *Finnegans Wake* may have of these books, the books themselves have more. There are 535 question marks in *A Portrait of the Artist as a Young Man*, 1,510 in *Finnegans Wake*, a whopping 2,235 in *Ulysses*. In each case the average number of question marks is just over two per page.[1] The number and pace of this stunning Irish Inquisition (as the act of counting has led me to think of the *Wake*, in particular) indicate, at the very least, that the text wants to know something about its reader. It seems that Joyce, whom César Abin caricatured, at his subject's own suggestion, as a question mark (Ellmann 645),[2] is intent not merely on telling a story to the reader but on asking of the reader – well, what, exactly? My argument, which holds together the following considerations of *Finnegans Wake* as 'a question,' is this: however persistently we may ask what the *Wake* 'is,' it demands of readers (and would-be readers, if there is a difference) with even greater polymorphous assiduity: 'What are you?'

There are several purposes and thematic relations to the repeated questioning. A 'question' posed as *the* question is a determinist device Joyce mimics ad absurdum ('the tonsure question' [FW 43.12–13]; 'the space question' [160.36]; 'Zot is the Quiztune' [110.14]; etc.). Joyce is ever-aware of its employment in the rhetoric of xenophobia (e.g., 'the Irish Question' or 'the Jewish Question,' the whole question of which might be expressed as 'what to do about *them*?'), and intrusion (e.g., judicial interrogation, inquisition by imposed and accusative authority).³ I will turn to the theme and structural value of questions after some brief consideration of the epistemology of the question mark.

Punctuation as a codified, standard-seeking system is in many ways itself the plainest proof of Derrida's contention that writing in western culture and tradition precedes speech, 'directing' as its constituent marks do the oral performance of a text. When one studies the importance of punctuation in Joyce, the history of its generation and usage becomes a central consideration, not least because it was the Irish who were, as M.B. Parkes reports, 'the first to develop certain new graphic conventions – features of representation and display – to facilitate access to the information transmitted in this "visible" medium [the written text]' (Parkes 23).⁴ Joyce, super-hibernophile sometimes in spite of himself, could hardly have been unaware of this history. Whether the 'numerous stabs and foliated gashes' and 'paper wounds' (FW 124.02–3) inflicted upon the pages of the *Wake* actually 'facilitate access' is an attractive conundrum. The text itself proposes that writing serves 'to = introdùce a notion of time [ùpon à plane (?) sù ' ' fàç'e'] by pùnct! ingh oles (sic) in iSpace?!' (124.10–2): this is partly just another disjointed rejoinder to Wyndham Lewis and his thesis in *Time and Western Man*, but the expression also slyly belies the idea of *faire violence au texte*, in that writing is itself a violent rupture, an attack upon 'plane' paper and in turn the reader's sense of vision and identity, 'iSpace.' The calmative reassurances of guidebooks notwithstanding, the reader who feels overcome by the *Wake's* orthographic onslaught needs to appreciate the vitality of this sensation. He or she may echo the book's own sense of estrangement and complain of its pages, 'how interquackeringly they rogated me' (FW 542.23–4). Question marks in the *Wake* function as signs of query (the Latin verb *rogare*, whence 'interrogate'), deceit ('rogue,' from the same root), and mortality (the Latin noun *rogus* means 'funeral pyre,' an inevitable sight at a wake), sometimes all at once. For Joyce, typography becomes typology.

The question mark is the result of an evolution in liturgical study, where corrective scribes were among the first to employ *positurae* like the *punctus interrogativus*. Such symbols' absorption into the act of writing – authors like 'Petrarch and Boccaccio paid close attention to the punctuation of their own works, drawing on the widest possible range of symbols available to them' (Parkes 48) – reflects the growth of textual awareness among early producers of texts, a trend that would produce experimentation in later centuries among the Scriblerians, the modernists, the language (or L = A = N = G = U = A = G = E) poets. In his useful study *Pause and Effect*, Parkes writes: 'The system of *positurae* was essentially a part of monastic culture ... The extension of the system, and particularly the correction of earlier books, reflects the dual rôle of the *precentor* or *armarius* in the monastery: various customaries indicate that he was responsible not only for the chant but also for the preservation and correction of books in the house' (Parkes 38). That the system which 'the company of the precentors and of the grammarians' (*FW* 26.21–2) invented for so-called corrective (more likely normative, revisionary) purposes is used for apparently de-harmonizing, chaotic ends in *Finnegans Wake* demonstrates part of the work's attitude towards attempt at textual 'correction.' The incorporation of revision as method in the *Wake* is represented as 'the revise mark' that 'stalks all over the page' (*FW* 121.02–3).[5] In this way the *Wake* is its own revisionist, shifting the words and letters around the constant marks, many of which look like this: ?

Proust remarks on 'ces phrases interrogatives de Beethoven, répétées indéfiniment, à intervalles égaux, et destinées – avec un luxe exagéré de préparations – à amener un nouveau motif, un changement de ton, une "rentrée"' (2:605: 'those questioning phrases of Beethoven's, indefinitely repeated at regular intervals – with an exaggerated lavishness of preparation – to introduce a new theme, a change of key, a "re-entry"' [Moncrieff and Kilmartin 2:627]). Joyce appreciated linguistic motifs as little different from those of music, and the *Wake*'s own 'phrases interrogatives' emerge and re-emerge transformed by the preceding pages' wordplay, but recognizable as variations. There may in fact be only a very few phrases, or phrase structures, within the book, which few Joyce stretches and bends, explodes and deflates. When variations of questions like 'when is a man not a man?' and 'was life worth living?' resound without definite answer, they persist in introducing 'un nouveau motif, un changement de ton, une "rentrée."' From this vantage point we perceive Joyce's punctuation as truly musical notation – however discordant or unusual – with the question mark as the 'ricocoursing' (*FW* 609.14) point of eternal return.

II

There is a literary tradition of interrogatory texts, the most obvious forms of which are the riddle and the catechism.[6] The sense of 'read' (from the Old English) as 'to interpret' is also found in the word 'riddle,' an important relation to bear in mind here. Joyce's riddling texts, especially *Finnegans Wake*, challenge, tease and mock us as we approach – but this is a critical path fairly well beaten. Margot Norris, for example, makes reference to the appearance of the riddles in Joyce's earlier works, such as Athy's Rumplestiltskin-like riddle upon his name (which can mysteriously be asked 'another way') to young Stephen, who admits he is '[n]ot very good' at riddles (*P* 25), and a more mature Stephen's 'hard' riddle about the fox given to his eager students in 'Nestor' (*U* 32). (Lenehan's spleen-poking 'Rose of Castille' / 'rows of cast steel' riddle [*U* 170] – with the unique Aeolian headline '? ? ?' [167] – can also be included in this collection.) Norris points out how these riddles are not answered correctly (92), except after the fact by the riddler himself. Probably the most intriguing riddle in *Ulysses* does not appear fully in the text and is not voiced by Stephen so much as it is simply thought:

Riddle me, riddle me, randy ro.
My father gave me seeds to sow.   (*U* 31)

These, Patrick McCarthy observes, are the opening lines to a traditional riddle to which the answer is 'writing' (36–7). In *Ulysses* writing is a riddling gesture; in the *Wake* reading involves endlessly interpreting so many riddles (recall that Joyce kept the *Wake*'s title a secret and riddled his friends about it: 'Tell your title?' [*FW* 501.02]).

Note that I say 'interpret' rather than 'answer.' '[D]efined in terms of its subject and meaning,' McCarthy admits that *Finnegans Wake* 'cannot be "answered"' (154); attempts for absolute meanings, like those of the producers of *Wake* summaries, end in an often ridiculously unsatisfactory way (think of Stephen's consternation over his inability to find 'the right answer to the question' of whether or not he kisses his mother every night before bed [*P* 14]). Of course, the riddle-without-answer is a pervasive motif in highly playful comic works. Examples range from the Mad Hatter's famous unexplained riddle of the raven and writing desk (Carroll 68–71) to this Marx Brothers dialogue from *Duck Soup*:

GROUCHO: Now what is it that has four pairs of pants, lives in Philadelphia, and it never rains but it pours?
CHICO: 'At's a good one. I give-a you three guesses.
GROUCHO: Now let me see ... has four pairs of pants, lives in Philadelphia ... is it male or female?
CHICO: No, I no think so.
GROUCHO: Is he dead?
CHICO: Who?
GROUCHO: I don't know. I give up!
CHICO: I give up, too.

No satisfaction: the jarring distinction from 'satisfy' is in the 'Nightletter' lines, 'To me or not to me. Satis thy quest on' (*FW* 269.19–20), with the recognition of questioning as questing.[7] McCarthy notes: 'Joyce's concept of the creative artist seems always to involve some form of riddle: the riddler is the equivalent of the Daedalian artificer, for the riddler is a form of verbal labyrinth whose purpose is to puzzle or mislead' (30). The 'masterbilker' (elsewhere 'monsterbilker' [296.07] and animus-counterpart to the mythical 'prankquean' [21.15], whose questions recur in alike groups of three) is this riddler. Hélène Cixous acutely attributes Joyce's manner to a 'writing governed by ruse,' which she finds is 'sometimes restrained, finely calculated, strategic, intending by the systematic use of networks of symbols and correspondences to impose a rigid grid on the reader, to produce an effect of mastery; sometimes, on the other hand, within the same textual web, surreptitiously, perversely, renouncing all demands, opening itself up without any resistance to the incongruous, introducing metaphors which never end, hypnotic and unanswerable riddles, a proliferation of false signs, of doors crafted without keys: in other words (spoken in jest), it is an extraordinarily free game' (19). Cixous observes that a key component of Joyce's game is his method of 'putting a question mark over the subject and the style of the subject' (15).* Thus, this transcendental question mark materializes over the head of the reader, both as the cartoon sign of bewilderment (reader to *Wake*: '?') and direct challenge to the reader's position as interpretive agent (*Wake* to reader: '???').

There are a few other relevant literary traditions of questioning besides catechism (question as instruction) and riddles (question as game). Philosophy actually has limited use for syllogism and typically presents propositions in a most provocative manner as a question.[8] Consider Aristotle's *Problems*. Probably collected as late as the fifth century B.C., it is a remarkable thirty-eight–book compendium of questions concerning a startling variety of subjects, including perspiration, hangovers, fear and courage, pleasant and unpleasant odours, mathematics, hot water, and justice.[9] Despite its use in philosophical writings and unlike its better-established cousins, the list and the litany, the quiz has not been given much focus in literary criticism. This state of affairs is especially odd when one considers the prominence the act of questioning has in related disciplines, especially linguistic studies. Aristotle himself significantly poses the question, 'Why are contentious disputations useful as a mental exercise?' within Book XVIII of *Problems*, 'Problems Connected with Literary Study' (1427). This format of philosophical writing begins to intersect with poetic directions after Wittgenstein. For example, Ron Silliman's 'Sunset Debris' is a poetic effort at sustained questioning, a kind of response to Wittgenstein's proposition-by-question form of investigation: 'Where do the words come from? What if we drained them of their meaning just to see what remained? What if we said we had done this thing? Can you give a yes or no answer? Can you say it in a few short words? How is it with all this language there is still this thing so vast that we have no name for it, even if we sense it as a thing we have seen? Were the words trapped in the pen, just waiting? Did they burst, sperm-like, into meaning in our mouths? Can you taste it? Can you feel it? What about it?' (40). By turns maddening, flirtatious, and ridiculous, 'Sunset Debris' is really a miniature of the Aristotelian contemplations and the Joycean onslaught.

The most basic and thus typically least thought of tradition of *Fragenkatalog* is the letter, in which one freely makes enquiries, however penetrating or banal, irrelevant or quotidian, of the addressee. (The addressee's absence effectively makes every letter the apostrophe missing from Joyce's title.) 'This letter must be full of questions' Joyce's mother sheepishly writes to her son in 1903 (*L II* 36), though it is nowhere near as 'full of questions' as the letters Joyce often sent to friends and relatives concerning often minute details of Dublin life, or, ultimately, the *Wake*, 'a letter to last a lifetime' (*FW* 211.22).

All of these forms – catechism, riddle, investigation, and letter – merge in Joyce's 'testies quoties questies' (*FW* 98.34). The quiz is a difficult quest for a reader, in both senses of 'for.' Joyce's favourite episode in *Ulysses*, 'Ithaca,' itself has 309 questions, by Richard E. Madtes's count (xi), and for this impertinence it has been the most critically spurned chapter in the book, with luminaries such as William Empson, Edmund Wilson, and Philip Toynbee included in the 'chorus of disapproval' (Madtes 65–6). Despite the great level of comedy within the chapter (the slapstick and the absurdity of home life's little details), 'Ithaca' unsettles and irritates many readers for its stylistic method. 'It is a method,' Madtes writes in a sentence that smartly coils itself into a question, 'evolving ultimately from the fundamental curiosity of inquisitive man in an incomprehensible universe, for what is Ithaca but the inevitable development of the first tentative, prehistoric "Why?" which cursed – and glorified – the consciousness of rational man? (67). These first gestures of enquiry, being the primal epiphanies, fascinate Joyce and recur in his work, from the earliest confrontation of *A Portrait* –

> And one day he had asked:
> – What is your name?
> Stephen had answered:
> – Stephen Dedalus.
> Then Nasty Roche had said:
> – What kind of a name is that? (*P* 8–9)

– to the primitive speech of the utterer and stutterer, Mutt and Jute: 'Scuse us, chorley guy! You tollerday donsk? N. You tolkatiff scowegian? Nn. You spigotty anglease? Nnn. You phonio saxo? Nnnn' (*FW* 16.05–7).[10] Wolfgang Iser comments that '[i]n the "Ithaca" chapter, aspects are not static but seem to be moving' (222): the shifting landscape, the unstill life there compels the reader to stabilize him- or herself, to hold fast to an Archimedean fixed point and observe. (This is impossible, but the attempt is nonetheless imperative.) The kind of 'negative capability' or vicarious presence a reader typically enjoys is nullified. Homer's 'Ithaca' is a homecoming in which Ulysses has to re-establish his identity as lord. Joyce spins the problem back at the reader: who are you, reader, stranger?

Naturally, *Finnegans Wake*, a perpetual linguistic motion machine, only exacerbates the 'Ithaca' reader's problem. The song from which Joyce cribbed his title ends on a primal question:

> Tim revives! See how he rises!
> Timothy rising from the bed!
> Crying, 'Whirl your whiskey around like blazes,
> Thanam o'n dhoul, do ye think I'm dead?'

Like everything else within his range of experience and observation, Joyce squeezed these words into the *Wake* – 'Anam muck an dhoul! Did you drink me doornail?' (*FW* 24.15) – with a difference. In 'Anam muck' one hears the animation in the primordial muck, and life expresses itself as a question. That the question-filled *Finnegans Wake* is itself in no sense 'dead' is the next link in my argument's chain.

III

More than any of its numerous and severe antagonists, the *Wake* reviews and questions its own methods and madness. It is not Joyce who laments, 'is there one who understands me? One in a thousand of years of the nights?' (*FW* 627.15–16), but 'the book of the opening of the mind' (258.31–2), *Finnegans Wake* itself. 'I quizzed you a quid (with for what?) and you went to the quod. But the world, mind, is, was and will be writing its own wrunes for ever, man' (*FW* 19.34–6). The interesting question here is: *whose* mind does the *Wake* have in mind?

The reader of the *Wake* again and again faces repeated probing and questioning related to the effort of reading. Do you understand all of this? Are you still awake? (Of all the questions on literary study in Aristotle's *Problems*, the one given most space concerns readers 'overcome by sleep even against their will' [1427–8].) And – are we having fun yet?

All a bit much, it might seem. Yet it was from his nagging wife, Stephen Dedalus asserts, that Socrates learned dialectics (*U* 243). Eglinton 'shrewdly' supposes that a bad marriage is a mistake, but not a 'useful' one.[11] On the contrary, the mistake of the apparently mismatched marriage, the very fount of *Exiles*, 'The Dead,' and of course *Ulysses*, with its lasting image of spouses sleeping head to toe, wrong only by dreary custom's measure. Joyce's idea of marital satisfaction might be summarized as knowing how and when to ask questions. When the Conroys return home from the Christmas party, Gabriel's asking of Gretta numerous questions (eighteen question marks there) is what leads him to his despair; he is more in the habit of making overpolished statements and is anxious when questioning.[12] By contrast, the 'yes' of the Blooms is the result of a consensual question, which need not even be fully articulated: 'I gave him all the pleasure I could leading him on till he asked me to say yes and I wouldnt answer first ... and then I asked him with my eyes to ask again yes and then he asked me would I yes to say yes' (*U* 932–3). The asking is continuous in *Finnegans Wake* – where 'ask' flickers to 'angkst' (*FW* 224.31) – because the reader's progress is not like a marriage, but like a seduction: 'him with his pregnant questions up to our past lives' (*FW* 438.11–12), 'atkings questions in barely and snakking svarewords' (*FW* 436.11–12). However, the simplicity and safety of 'yes' and 'no' answers are absent from the interrogatory strategies of the *Wake*. This book questions its reader, 'suspecting the answer know' while 'expecting the answer guess' (*FW* 286.26–8). *Finnegans Wake* is a consciousness seeking another, perhaps greater consciousness (that of the 'ideal reader').

While it may first seem strange or implausible to refer to a text's consciousness, or simulacrum of consciousness, the concept of a poem or especially a novel as a container or impressed mould of consciousness is by no means new. Before the streams of Woolf and Joyce there is a lineage of vessels, from Don Quixote's basin-helmet to James's *Golden Bowl*, which holds in shape the respective madnesses and mischiefs that govern those narratives. The associational meanings of the *Wake*, Hydra heads that only multiply when attacked, reduce the degree of metaphor in the expression of 'a text's consciousness,' and in turn, startlingly, enhance that of metonymy.

George G. Colomb and Mark Turner point out that 'AI [the popular acronym for artificial intelligence], textlinguistics, cognitive psychology, *and* literary theory unknowingly share a body of questions and answers about the nature of meaning and understanding,' and they argue that 'AI, which has tried to conceive meaning too narrowly, will find that what literary theorists know about meaning is crucial to the agenda for future AI research. We believe that literary theory, which has tried to find value in "literary" meaning by marginalizing it, can learn from AI (and related inquiries) not only information essential for any study of meaning but also a new – and far more accurate – sense of the place of theory of literature in the general arena of textual studies' (388; italics in original). Mind, machine, and text are very distinct entities, but as independent semiotic systems they have useful points of comparison by which the greater mysteries of each may be better understood in connection with certain lesser mysteries of the others. This having been said, however, doubts about the conceit of the human mind as a machine, often implicit (the conceit, not the doubts) in studies of AI and cognition, are important and deserve greater attention. For my argument as it appears here, I urge my reader to question whether this popular conceit is any more valid than comparing the human mind to any other 'multi-tasking' human-made device (such as, for instance, a complicated text). Certainly, my claims for *Finnegans Wake* as an ('artificial') intelligence, or as detector of consciousness, draw in no small part on a metaphorical understanding, and I would not posit Joyce's book as anything exemplifying what specialists call 'true' AI. At the very least, though, I suggest that works such as Silliman's 'Sunset Debris' and Georges Perec's 1968 German radio play, *Die Maschine*, and *Finnegans Wake* play between what appears to be 'mechanical,' pounding repetition, and what Samuel Beckett calls the 'psychological inevitability' with which words expand ('Dante ... Bruno. Vico ... Joyce' 11) to such an extent that they demonstrate an associative capability at once independent of authorial 'intention' or presence and interactive with a reader's own associations.

Jean-Michel Rabaté's discussion of *Finnegans Wake* as 'an autonomous, almost automatic machine' (81; See also Lorraine Weir's relevant notion of the *Wake* as a 'McCarthy machine' [93–8]) is not all that radical when one considers that as early as 1955 Hugh Kenner was already thinking of *Ulysses* as a 'huge and intricate machine clinking and whirring for eighteen hours' (*Dublin's Joyce* 166), but the difference between these 'machines' is significant. The encyclopedic quality of Joyce's work, the understanding of *Ulysses* as archive, is at a general level now a critical truism. Yet where *Ulysses* stands as a record aptly mimicking life, a meticulously programmed virtual Dublin day, the density of multilingual information within a lesser space (it always seems strange to note that the *Wake* is actually shorter than its predecessor in all editions) is too volatile for a repeat performance in the *Wake*. Instead, a faster, noisier, far more entropic entity blinks back at its would-be operators, always ready to 'Asky, asky, asky!' (FW 233.27). Kenner argues: 'In rejecting the hylomorphic doctrine that things are intrinsically intelligible, post-Cartesian philosophy placed itself in H. C. Earwicker's posture of "suspensive exanimation," producing by a twist of the hand an infinite succession of private geometrically-ordered worlds ... the pseudometaphysics of Malebranche, Descartes, Hume, Kant, and Locke and the phenomonological dazzle of Times Square' (*Dublin's Joyce* 316). More than a database, the *Wake* is, as I said in an earlier chapter, an indigestible digest, because it is data in digestion. Information is accessible only in process – though this is exactly the principle that rescinds the false guarantee of the information's intelligibility. When Jed Rasula writes of the *Wake*'s 'archival mass as a rubbish heap fermenting provocative incitements that do not so much illuminate as thicken or increase the texture of the darkness' ('Indigence in the Archive' 37), the active word is 'fermenting': Joyce recedes from the pub taps of *Ulysses* to meditate on the alchemical, patient act of brewing. Of the intriguing phrase 'states of suspensive exanimation' (*FW* 143.08–9), which Kenner mentions, two important associations now materialize: the conflation of sustainment with suspension is felt in Barthes;[13] the resurrection (waking) comes of interrogation ('exagmination'). This is going to be a long quiz, because we as readers are such heavy sleepers, trapped in the nightmare of history.

In case 'cogito ergo sum' still requires any refutation, it turns out that even a dumb book can parrot Descartes's jingoism: 'cog it out, here goes a sum' (*FW* 304.31). Jacques Maritain, a philosopher Joyce enjoyed, considered that 'Cartesian thinkers ... imagine or construe the object as a piece of reified externality, dead, an affront to the mind until the mind has processed it' (Kenner, *Dublin's Joyce* 135). Exchanging 'text' for 'object' will reveal 'Cartesian thinkers' (rather like Berkeleyan thinkers) in this context to be equatable with many critics, particularly – though by no means exclusively – adherents of reader-response theories. In fact, literary criticism as a whole now struggles with logocentrism while generally leaving unchecked its lectorcentrism, if that is not altogether too ridiculous a neologism. The assumption that the reader's agency is in some fashion independent and central to interpretation is in modern contradistinction to classical thought, and it has been buttressed by the dominant discourse of identity politics in current cultural criticism. Wittgenstein glares coldly at this kind of assumption – 'Die Deutungen allein bestimmen die Bedeutung nicht' ('Interpretations by themselves do not determine meaning' [80]) – which is what separates him from other thinkers on interpretation, such as Wolfgang Iser and Stanley Fish.[14] Despite more and more recognitions of any given text's individual history, the notion of a text as a living thing is rejected not as just another musty liberal humanist cliché but very much as 'an affront to the mind.' Flattery has quietly shifted away from the idea of author, or 'author function,' to the idea of reader, without losing its intensity.[15] The *Wake* enjoys reminding readers that they are too complacent and self-congratulatory ('As any explanations here are probably above your understandings' [*FW* 152.04–5]), invites them to generate contexts for the flow of discombobulated words and phrases, and casts doubt on the reader's sensory perceptions: 'The mixer, accordingly, was bluntly broached, and in the best basel to boot, as to whether he was one of those lusty cocks for whom the audible-visiblegnosible-edible world existed. That he was only too cognitively conatively cogitabundantly sure of it because, living, loving, breathing and sleeping morphomelosophopancreates, as he most significantly did, whenever he thought he heard he saw he felt he made a bell clipperclipperclipperclipper' (*FW* 88.04–11). Shapes of Berkeley, Socrates, Pavlov, and an ectoplasmic Oscar Wilde[16] all can be traced in these lines: each expresses a philosophy based on interpretation. It is Barthes's libertine reader who is being cross-examined here. Joyce's riddling machine is attempting to attain its own jouissance, or 'joyance' (*FW* 598.25).

Edgar Allan Poe's orang-utang, seeking to emulate with politesse and the accoutrements of a human (those same items, the mirror and razor, with which *Ulysses* begins), butchered women in the Rue Morgue. There is no comforting 'nevertheless' in that proposition. James Joyce fed his supercomputer as much compressed knowledge of human experience as he could, and now, understandably, it has a few questions.[17] Such monstrous, imitative intelligences should be carefully attended.

IV

How many of Joyce's questions are answered? Ultimately, not many – at least, not to the point of closing the question. Wrong answers, insincere answers, inexplicable answers abound.[18] Stephen's fox riddle is almost comprehensible when compared with the strange distance between the questions of the 'nightly quisquiquock of the twelve apostrophes' (*FW* 126.06–7). For the same emphasis upon incoherence there is also the use of indelicate non-answers, such as the aforementioned 'N ... Nn ... Nnn ... Nnnn' sequence (*FW* 16.06–7) and the more lively example, 'Bum!' (102.36). Sometimes the non-answer paradoxically acts *as* the answer, or a kind of prompt to the reader to essay an answer of his or her own. One of these potential replies to 'the first riddle of the universe' (170.04), namely, 'when is a man not a man?' (170.05), is 'WHEN THE ANSWERER IS A LEMAN' (302R01–3). This answer is neither 'when the answerer is a man' (which might render the question moot), nor the sour 'when the answer is a lemon' (McCarthy 98; see *U* 575), but an uncategorizable hybrid ('Miscegenations on miscegenations' [*FW* 18.20], indeed), just another question for the reader. 'No answer' appears repeatedly in the *Wake*, though only as distortions, such as 'Noanswa' (23.20), 'Nuancee,' and 'Noahnsy' (105.14). In Gaelic, Patrick McCarthy reports, 'ni ansa' means 'not hard (to say)' (30), while

Learner's Irish-English Dictionary defines 'ansa' as 'preferred: more (most) loved' and 'ní' as 'thing.' The sensical reader would prefer an answer and, like Alice at the Mad Tea Party, cannot fathom the value(s) of 'asking riddles that have no answers' (Carroll 71).

Such values exist, however, and are extremely important. Simply to posit that questions without answers exist is itself an affront to reason, but to do so at such length and simultaneously to tease with the possibility of answers after all '(for teasers only)' (*FW* 284.16) demonstrate a determined attack upon inflexibilities in reading and understanding. A practical example of an operation that employs the madcap method of nonsensical catechism is the Turing Test, the sustained interrogation process by which claims of artificial intelligence are put to task. In the test, a human interrogator submits questions to a pair of unseen test subjects: one is a human and the other the computer allegedly programmed to think like a human being. The subjects must promptly answer each question, sequentially.

> For whatever question one might first suggest, it would be an easy matter, subsequently, to think of a way to make the computer answer that *particular* question as a person might. But any lack of real understanding on the part of the computer would be likely to become evident with *sustained* questioning, and especially with questions of an original nature and requiring some real understanding. The skill of the interrogator would partly lie in being able to devise such original forms of question, to see if the computer could detect the difference, or she might add one or two which sounded superficially like nonsense, but really did make some kind of sense: for example she might say, 'I hear that a rhinoceros flew along the Mississippi in a pink balloon, this morning. What do you make of that?' (Penrose 9–10; italics in original)

Within the parameters of the Turing Test principles, the ability to judge or determine or self-consciously create 'nonsense' (in this case, critical nonsense) is a significantly (read: particularly and perhaps exclusively) human attribute.

The necessity of a '*sustained* questioning,' stressed by Roger Penrose in his description of the test, is likewise an integral part of the questions and nonsense of *Finnegans Wake*. Like the combinations of genetic nonsense that form different human beings (and, of course, all living organisms), the possible differential mutations upon polylinguistic and textual signifiers (including punctuation) are innumerable, and reading them would take 'for ever and a night,' as the *Wake* waggishly claims to require (*FW* 120.12-13). If we think of *Finnegans Wake* as a sort of literary Turing Test, a trial-and-error routine by which humanity, or humanness, can be differentiated from an artificial construct, we can appreciate the zealous use of questions; for then we can clearly 'hear the riddles between the robot in his dress circular and the gagster in the rogues' gallery' (*FW* 219.22-3).

Peter Szondi's essay 'Über philologische Erkenntnis' is known in English as 'On Textual Understanding,' though its title also could be translated fairly as 'On Literary Cognition.' Szondi contemplates literary study's claims to be a science and grants that this claim is tenable 'only if [such study] immerses itself in the works themselves,' stressing Adorno's logic of 'a productive process' (22). In the course of this essay Szondi notes a discouraging shift in the epistemology (another possible translation for *Erkenntnis*) of literary study: 'The activity through which knowledge is enriched and transformed is called "research" *Forschung*) ... *Forschen* once meant "questioning" and "searching," as in the expression "an inquiring look" (*ein forschender Blick*). But the element of questioning, and thus also of understanding (*Erkenntnis* [or 'cognition']), implied by this word has steadily diminished, and research has become simply a matter of looking for items of knowledge. From the very way in which he speaks of his "research projects," the literary scholar admits that he views his efforts as consisting more in seeking out something that exists and that it is his job to discover than in cognizing and understanding' (6–7). The interrogation of *Finnegans Wake* presents a vortex of phenomenology and aesthetics in its own unblinking 'inquiring look': 'Fas est das and foe err you' (*FW* 273.06). The deluge of questions is an opportunity for the reader, even the literary scholar, to recognize his or her own cognitive abilities. It is a chance to test one's own humanity, errors and all.

Erroneous Conclusions

Having now been 'subjected to the horrors of the premier terror of Errorland,' my reader may '(perorhaps!)' (*FW* 62.24–5) wonder whether a text's reader or its author has a greater potential – or more right – to be dissatisfied with it. Ten years after the publication of *Finnegans Wake*, Samuel Beckett submitted to Georges Duthuit that 'to be an artist is to fail': 'I know that all that is required now, in order to bring even this horrible matter to an acceptable conclusion, is to make of this submission, this admission, this fidelity to failure, a new occasion, a new term of relation, and of the act which, unable to act, obliged to act, he makes, an expressive act, even if only of itself, of its impossibility, of its obligation' (*Three Dialogues* 125). Speaking here is the Beckett who learned from Joyce, and it is striking how akin (though ever so much more sharply honed) these sentiments are to the salutations of Stephen Dedalus: 'I am not afraid to make a mistake, even a great mistake, a lifelong mistake and perhaps as long as eternity too' (*P* 247). Beckett's impossible obligation to art holds true for criticism, too. Joyce's aesthetic of error, as measured in the previous chapters, demands participation, with the same urgency with which a virus seeks a body to infect. The (sic)ness that is Joyce's contribution to the literary corpus and to the 'textual condition' is not to be suppressed by clinical application of academic discourse, because no discourse that engages the texts of Joyce is immune. (Please see the appendix for brief case studies supplementary to those discussed previously.)

In my introduction I proposed a critical framework, called for by writers such as Groden, in which a balance would be struck between theoretical concerns and a rigorous awareness of textuality. In the course of the attempt – the 'success' of which is not for me to judge here, though I will spare a last moment to contemplate its failure – the resultant discourse has revealed a second, equally important agenda. Walter Benjamin's advocacy of a criticism of quotational collage (*Das Passagenwerk* is his insurmountable exemplar) has long interested me, and the possibilities of a fusion of scrupulous use of this collage method with the volatile 'hides and hints and misses in prints' of Joyce and Joyce criticism seemed risky (that is, more than merely 'problematic') and exciting. Would a collage of *misquotation*, a *structure of mistakes*, hold together or collapse? This is not really the question; of course, the centre will not hold. My early emphasis on theory as 'probe,' after McLuhan, was expressly predicated upon the understanding that any such 'probe' is interesting only when on its voyage it faces interference, registers anomalies, is confronted, or is actually destroyed. Only in turbulence and adversity, only in its failure does a probe contribute meaningfully to the understanding of its designer(s). The newspaper that reported that 'Pioneer 10 carries a message ... in the form of a plague designed to show ... the place and time where it began its long journey' (Hobbs 11) reveals the organic and interactive (or at least reactive) nature of such a probe, and the 'plague,' by dint of its erroneousness, is indeed a measurement of the crossing of space and time.

Those transmissional processes or acts, the *attempts* at which I have called here simply 'writing' and 'reading,' are certainly cognitive wonders, but 'unless we were never so wrongtaken' (*FW* 586.31-2), it is their lapses that make them individual and instructive and even artful. Without error we would, I think, have none of these human qualities. Modernism's and especially Joyce's textual fault lines underline this concept and offer it as an undervalued factor in formulations of intelligence and its manifestations. In a 1918 essay entitled 'On the Program of the Coming Philosophy,' Benjamin sketches outlines for a revision of the Kantian system of thought. In this important and *modern* revision the priority is 'to determine the true criteria for differentiating between the values of the various types of consciousness' (104). Benjamin writes: 'with the radical elimination of all those elements in epistemology that provide the concealed answer to the concealed question about the origins of knowledge, the great problem of the false or of error is opened up, whose logical structure and order must be ascertained just like those of the true. Error can no longer be explained in terms of erring, any more than the true can be explained in terms of correct understanding' (107). *Finnegans Wake*, the culmination of Joyce's experiment of literature in error, actively shakes up the expressions themselves of just 'those elements of epistemology.' 'Epistlemadethemology' (*FW* 374.17) emerges and deviates from hermeneutic norms and refuses to distinguish information or allegation from its transmissional ups and downs, including those of the present reading. Octavio Paz considers the twentieth century to be 'the century of the return, by unsuspected paths, of a power denied or at least disdained since the Renaissance: the old inspiration. Language creates the poet, and only in proportion as words are born, die, and are reborn within him is he in turn a creator. The vastest and most powerful poetic work in modern literature is perhaps that of Joyce; its theme is immense and exiguous ... The poem devours the poet' (255). And, I would add, the reader, too. Joyce's texts return us to the smithy of our souls (*A Portrait*), the home of the 'incomplete' (*Ulysses*), 'Dublire, per Neuropaths' (*Finnegans Wake* 488.26): the cerebral flashpoint at which expressive opportunities for error and errancy begin.

That the imperfections of my argument *are* my argument will and should not please everyone. Certainly, I have had occasion to sidestep a direct critical path (a construction that may very well be an inadequate fiction), even for the sake of a pun, and though I have striven to be faithful in reproducing and representing Joyce's texts, I am aware that I, too, have been subject and agent to the inevitable process of distortion that I have outlined in the preceding chapters. But as Clive Hart has had occasion to note, 'total clarity of recall is by no means incompatible with total error. It is of such stuff that literary history is made' (10).

The greatest problem with this book – or at least, so it seems to me now, and again, not that I am a valid judge – is the false act I seek to perform at this moment. Errorland has no exit. Blanchot remarks that

> dans cette région qui est celle d'erreur parce qu'on n'y fait rien qu'errer sans fin, subsiste une tension, la possibilité même d'errer, d'aller jusqu'au bout de l'erreur, de se rapprocher de son terme, de transformer ce qui est un cheminement sans but dans la certitude du but sans chemin. (Blanchot 92)

> (in [the region of error] one does nothing but stray without end, [and] there subsists a tension: the very possibility of erring, of going all the way to the end of error, of nearing its limit, of transforming wayfaring without any goal into the certitude of the goal without any way there. [Smock 77])

As 'one more unlooked for conclusion leaped at' (FW 108.32–3) regrettably approaches, like the ground hastening to greet one who tries to fly by certain nets, I hope that my reader shares with me this comfort: the moving blots that are Joyce's mistakes make our responses as readers and writers more 'human, erring and condonable' (FW 58.19).

APPENDIX

Quashed Quotatoes

> Note the notes of admiration! See the signs of suspicion! Count the hemisemidemicolons! Screamer caps and invented gommas, quoites puntlost, forced to farce!
>
> (FW 374.08–11)

Hanging's too good for a man who makes puns, the comedian Fred Allen once quipped: no, he should be drawn and quoted. Joyce's ambivalence towards quotation is fundamental to his uses of and developments in language and narrative. In a 1903 review of A.S. Canning's *Shakespeare Studied in Eight Plays*, a twenty-one-year-old James Joyce complains that quotations from the plays 'fill up perhaps a third of the book' and that the treatment of those quotations is 'remarkably irreverent' (*CW* 137). 'It is not easy to discover in the book any matter for praise': speeches are abbreviated, and Joyce bristles at 'misquotations' – despite the fact that he, in turn, misquotes Canning, writing 'fully' in his review where should be instead, of all possible words, the name 'Ulysses' (137). He concludes sourly, 'even the pages are wrongly numbered' (138).

Joyce's awareness of the regularity of misquotations (his own as well as others) within the 'textual condition' becomes a dramatic principle in his fiction. Fogarty misquotes Dryden in 'Grace' (*D* 168); Stephen misremembers a line from Thomas Nashe (*P* 234); Deasy quotes *Othello* without regard for context (*U* 37); Best fumbles over whether 'thing of beauty' is what 'Yeats says' or ' Keats says' (*U* 627). (Pedants, all four quoters, one notices.) The vulgarity of these gestures, as it were, may well lend them the status of epiphanies, at least as Joyce understood them. Shakespeare remains the favourite hobbyhorse, the central canonical figure whose words are ostensibly cherished but mangled in forgetful or pretentious mouths. Bloom comically sees in Shakespeare a fount for advertising slogans: 'Music hath charms Shakespeare said. Quotations every day in the year. To be or not to be. Wisdom while you wait' (*U* 361).

The (sic)ness of Joyce's texts, which I have attempted to outline in the preceding chapters, becomes a contagion in criticism of his works; for those works are almost inevitably misquoted. Eloise Knowlton complains that '[w]hen we cite, we cite Joyce. Himself' (63), that we as readers are helpless to do anything but parrot the master of language. Actually, Joyce's use – or misuse – of Shakespeare itself represents the counterargument. When we cite Shakespeare or Joyce, we mistake them, remake and rewrite them in error.

Thus, a newspaper column may reduce *Ulysses* to this description (a failed attempt at parody): '[a] novel about three Irish men and one of their wives who spend three days teaching, drinking, talking, shaving and complaining about certain aspects of their life and culture. Also explores female sexuality' (Jackson D22). Three men? (I assume Buck Mulligan is counted, given the 'shaving.') Three days? As Fritz Senn might say, this *Ulysses* is not the book I know – or the book I think I know.

Below is a short but representative list of instances of Joyce misquotation in effect. The numbers on the left refer to the page numbers in this volume where the corresponding works and/or authors are discussed or cited.

42   Vogler's essay has an interesting error in its parenthetical citation for a pair of epigraphs from Joyce. The first is from '*U* 9.228–9' all right, but the next is mapped at the impossible co-ordinates '*FW* 9.228–9' (Vogler 201): a printer's stammer. The *Wake* lines 'you may be as practical as is predicable but you must have the proper sort of accident to meet that kind of being with a difference' appear at *FW* 269.13–15: Vogler reprints it as 'to meet with that kind.'

76   In *Here Comes Everybody*, Anthony Burgess makes a few gaffes. In referring to one 'Emma Cleary' (53) in *A Portrait*, Burgess, like many others after him, assumes she is the character from *Stephen Hero* (she 'appears only in her ini-

tials, a cipher at the head of the poem [Stephen] writes for her' in *A Portrait*), but in that unfinished draft her last name is spelled 'Clery.' Another name problem: 'M'Coy' of both 'Grace' and *Ulysses* becomes 'McCoy' (Burgess 110). Also, Burgess strangely breaks up one of Joyce's compounds by producing 'all Ireland is washed by the Gulf Stream' (Burgess 112) for 'All Ireland is washed by the gulfstream' (*U* 18).

77 The *Norton Anthology*'s *Wake* excerpt has two other differences from the Viking text. The period after 'Forgivemequick' (2312, line 13) ought to be a comma, and the hyphen in 'daughter-sons' (2313, line 11) is contestable: although in the Viking edition of *Finnegans Wake* one finds the hyphen splicing the word for a line break, it seems unJoycean to break up a perfectly good portmanteau in this way.

102 An odd instance not of a misquotation of Joyce, but of a self-misquotation by a Joycean: in *The Finnegans Wake Experience*, McHugh transcribes an exegetical discussion of 338.22–7, including participants such as Clive Hart, Fritz Senn, and Rosa Maria Bosinelli. At one point McHugh records himself as having said, 'I don't think much if this is Chinese' (66), and I have to assume that 'if' should be 'of.'

106 John Bishop's *Joyce's Book of the Dark* reproduces 'intrepidation of dreams' instead of 'intrepidation of our dreams' (338.29–30).

107 *Our Exagmination* enjoys its healthy share of typographical flubs and inconsistencies. (For example, William Carlos Williams refers to a story in *Dubliners* called 'A Sad Case' [175].) In 1939, New Directions assumed publishing rights for the volume and promptly printed their first edition, but while they added an introduction by Sylvia Beach as late as 1961 for their second edition, they left the troubled text intact.

110 Eric McLuhan, in *The Role of Thunder in* Finnegans Wake, proposes a reading of 314.10–12 as a 'paraphrase' of dia-

	logue from the 'Emmaeus chapter of *Ulysses*' (164). McLuhan's recognition of a one-two-three sequence of identified speakers partly depends on the line '–Neat bit of work, longshoreman one said' (*U* 730), which appears as 'one longshoreman said' in post-1960 editions (*U-C* 660, *U-RE* 547).
138	Cixous in *The Exile of James Joyce* takes a number of liberties with Joyce's texts. The concluding cry of 'Counterparts' is quoted as 'I'll say an Ave for you' (21) instead of 'I'll say a *Hail Mary* for you' (*D* 94). Elsewhere Cixous writes that 'Stephen declares "the mistakes of genius are voluntary," justifying after the event with Satanic pride certain errors of his own' (xiv), when what Stephen actually says in *Ulysses* is that '[a] man of genius makes no mistakes. His errors are volitional and are the portals of discovery' (*U* 243).
170n1	In *Lars Porsena* Robert Graves makes the strange argument that '[t]he only character in [*Ulysses*] with whom [Stephen] Daedalus [sic] has a strong natural sympathy is his father, the only one man who is able to harmonize religion, politics, and obscenity into something like an artistic reality' (92).

Notes

1: Re: Cognizing Error

1 The cautions of Alan Sokal and Jean Bricmont on the 'abuse of science,' however overstated they sometimes appear, are to be observed here (see especially Sokal and Bricmont 183–9). My briefly sketched paradigms of critical attitudes towards intentionalism, which appear later in this chapter (e.g., comparisons with a 'Newtonian' world view), are only analogies and do not imply that the 'progress' of literary criticism, such as it is, by any means occurs in the same pattern or at the same pace as that of physics. As always I defer to the *Wake*: 'Sifted science will do your arts good' (*FW* 440.19–20).

2 José Ortega y Gasset offers a complementary view: 'Present-day science would be impossible without language, not because of the cliché that to produce science is to speak, but, on the contrary, because language is the original science. Precisely because this is a fact, modern science lives in a perpetual dispute with language' (105).

3 Bacon's use of the tropes of light, darkness, and caves belongs to the poetic language of the Renaissance. Besides the obvious echoes of the King James version of the New Testament, the reader may recall Spenser's dragon, Error: 'For light she hated as the deadly bale, / Ay wont in desert darkness to remaine, / Where plaine none might her see, nor she see any plaine' (1.16.7–9).

4 Unfortunately for Galileo, the publication of these and other ideas raised the ire of the Catholic church, which in 1616 had decreed such thinking 'false and erroneous.' For his 'errors,' Galileo was brought before the Inquisition.

5 This quip is not included in Hawking's unexpected bestseller, *A Brief*

History of Time, but rather is made in Errol Morris's documentary of the same title.

6 My insincere use of the poor word 'perceived' actually betrays my borrowing on Aquinas's '*visa.*' Cf. *A Portrait*: 'He uses the word *visa*, said Stephen, to cover esthetic apprehensions of all kinds, whether through sight or hearing or through any other avenue of apprehension. This word, though it is vague, is clear enough to keep away good and evil which excite desire and loathing. It means certainly a stasis and not a kinesis' (*P* 207–8). See the discussion in chapter 7 concerning 'the Dogberry Question.'

7 Although in a later chapter there is consideration of Pound's translation technique in the construction of a view of modernism's errancies, specific problems of translation are decentralized here, since the subject constitutes not just another study but an entire body of enquiry.

8 For clarification: the separation of syntactic from calculative errors is their different degrees of arbitrariness in their rules and principles (i.e., how logically demonstrable they are); their similarity lies in the shared arbitrariness of their notation.

2: The true scholastic stink

1 Certainly, there is a good feminist-psychoanalytic argument to be made about these 'portals.' Both of Stephen's discussions circle around but do not confront images and ideas of womanhood, particularly motherhood (the cow, the mother of Shakespeare's offspring, and May Dedalus), in each case framed with violence and death.

2 Patrick McGee, considering Hugh Kenner's 'system of limits' gloss on 'style' (*Joyce's Voices* 81) writes: 'T.S. Eliot read ['Oxen of the Sun'] as an expression of "the futility of all the English styles," a view corroborated by Bloom's Wildean sense of "useless words." The history that Joyce recapitulates in "Oxen of the Sun" is a history of metaphorical truth or a genealogy of styles that identifies the fecundity of writing with its sterility and wastefulness' (*Paperspace* 100).

3 To be strictly accurate (and to adhere to the principles of textual approach I am advocating here) this counting does not actually reflect the differences of textual appearance of these words. For the sake of this minor and easily demonstrated point about recurrences of word-types, I have not merely focused on homonymic qualities and recognizable English words; in this case I am looking only for the sequence of letters appearing as single words, without observing typographical nuances such as capitalization (e.g., 'air' appears thirty-two times, 'Air' once) or italicization ('Ere' ten, '*Ere*' once).

4 More on the 'untamed jungle' trope: McHugh notes that the 'original blurb sent out by the editors of *The James Joyce Archive* ... lists sixteen volumes of facsimiles of *Finnegans Wake* notebooks, which, it says, "constitute the dark continent of *Wake* studies"' (*The* Finnegans Wake *Experience* 82; ellipsis added).

Knowlton's book, squinting as it does at the authority of citation, has its own textual tic in familiarly abbreviating Joyce's title as '*FW*' even in the body of her arguments (just as McHugh does, incidentally), which is as annoying as the habit – at work in Robert M. Adams's *Surface and Symbol*, for example – of rearticling Joyce's first novel as 'the *Portrait*.' (Lorraine Weir flattens the roundest characters of *Ulysses* to algebraic designations, straight lines between points: 'SD,' 'LB,' and 'MB' [42–52].) Athough it may be a lesser transgression, signification of 'the *Wake*' is also a distortion, of which I am as guilty as anybody, but I think this construction reveals something of the book's unique nature. In some ways this distinction, like Borges's manner of referring to Cervantes's novel as 'the *Quixote*,' is a commendatory gesture.

5 After conceding that her own critical discourse is as 'quotational' as that of any other Joycean, Knowlton writes: 'Criticism's chances for substantive change depend on its ability to revise its constituting rhetoric, but in an era when the humanities are under seige from conservative critics and downsizing deans, this seems an unlikely moment to see such a shift' (113). My own contrary expectations are embodied in my preservation here of the spelling, 'seige.' Revising rhetoric is all well and good, but textual forces constitute the ultimate revisions.

6 See Conley, '"Oh me none onsens!" *Finnegans Wake* and the Negation of Meaning.'

7 A. Walton Litz has a negative view of this 'inherent defect in Joyce's method': 'Too often the process of deformation diffuses the basic effect instead of intensifying it' (92).

8 Cf. Terry Eagleton, *Literary Theory: An Introduction* (Minneapolis: U of Minnesota P, 1983) 82. Fritz Senn similarly calls the *Wake* 'the superlative' (*Inductive Scrutinies* 227).

9 Norris, however, ascribes the perception of 'errors' in the book to '[n]ovices to *Finnegans Wake*' and roughly limits the significance of 'misspellings, nonsense words, [and] malapropisms' to Freudian indications of 'unconscious truths' (*Decentered Universe* 101). To wit: 'error is caused by the repression of the writer's true feelings and their unconscious eruption in the misprint' (115).

10 A parallel study to which I refer the reader interested in these questions is

Mark C. Taylor's *Erring: A Postmodern A/theology*. Although Taylor's analysis is, as his subtitle suggests, largely theological, and mine is considerably more secular in its phenomenological underpinnings, I think we (with Deleuze and Guattari) share an interest in what he calls 'Nomad Thought': '[t]he erring nomad neither looks back to an absolute beginning nor ahead to an ultimate end. His [?] writing, therefore, remains unfinished. His work is less a complete book than an open (perhaps broken) text that never really begins or actually ends' (13).

11 See the appendix.

3: Fault Lines: Representing Modernism's Errors

1 Of course, the modernist phenomenon of unfinished work(s) and masterworks(s) is not limited to literature – one need only think of Tatlin's Monument to the Third International, or Schoenberg's *Moses und Aron* – any more than it is to artistic efforts. All of the best-laid plans, including Nazism's thousand-year Reich and *die Endlösung*, were not to be fully realized.

2 Joyce by this time has already postulated 'a limbo of painless patient consciousness through which souls of mathematicians might wander, projecting long slender fabrics from plane to plane of ever rarer and paler twilight, radiating swift eddies to the last verges of a universe ever vaster, farther and more impalpable' (*P* 191). Of special note in this passage is the coincidence of two tropes relevant to later discussion: wandering (the etymology of err) and fabrics (the weaver's effort).

3 Dismayed that 'Gödel's theorem is an inexhaustible source of intellectual abuses,' Sokal and Bricmont flatly deny any 'logical relationship between this theorem and questions of sociology' (176–7). The concern is valid, but I cannot sympathize with the narrowness of the rejection's expression (who says such a relationship must be logical?). In an assessment of Joyce's use of ideas of relativity in the *Wake*, Andrzej Duszenko, with admirable caution, points to the author's 'accretive' method; the scientific advancements were at least cursorily scrutinized and then selectively borrowed from: 'they were meant to enrich the texture of the book by providing new points of view, or commenting on the text, or even contradicting it to bring out its meaning' (62). The sillier commentaries on the 'scientific' qualities or structure of Joyce's works – and they do exist – are typically the result of ruminating on the matter with either much greater or less avidity than Joyce himself did.

4 Sartre's French text omits the final or meta-parenthesis, but I have added it here.

5 In his prefatory note to his reader, Browne refers to his *Pseudoxia Epidemica* as a *'Labyrinth'* (167), an apt characterization of its myriad-mindedness. Browne's pedagogical project can be appreciated as proto-modernist in its apologetics (*'we crave exceeding pardon in the audacity of the Attempt; humbly acknowledging a work of such concernment unto Truth, and difficulty in it self, did well deserve the conjunction of many heads'* [165; italics in original]) and the understanding expressed that the catalogue of errors is not a volume of truth.
6 In 'Ithaca' Bloom enumerates the 'imperfections in a perfect day' (*U* 860).
7 In his 1947 study *Call Me Ishmael*, Charles Olson remarks upon the multiplicity of Melville's texts as a struggle to contain the novel's subject: '*Moby-Dick* was two books written between February, 1850 and August, 1851. The first book did not contain Ahab ... It may not, except incidentally, have contained Moby-Dick' (39).
8 Whitman can be thought of as either a transition or a more general connective between Melville's view of text as monstrous and Moore's thinning verses. A good history of the reading of his accretive and multi-editional writing is Mike Feehan's 'Multiple Editorial Horizons of *Leaves of Grass*' (*Texts and Textuality: Textual Instability, Theory, and Interpretation*, ed. Philip Cohen [New York: Garland, 1997], 161–82).
9 Kappel's suggestion, however, that the additive/subtractive distinction is 'a difference of temperament between novelists and poets, or a difference between the modernism of the novelists and the modernism of the poets' (126) is jejune.
10 Postmodern authors enjoy the technology that allows such transformations to transcend the simply figurative. *Agrippa: Book of the Dead*, William Gibson's elegy for his father, is a notorious example. In interview Gibson describes the poem as 'a work of art that never existed. That was not my intention ... In a funny way, the thing never existed. It's a ghost, or again, it's an urban legend. The intention was that the text would be on disk (this all sounds so primitive), a disk that would have some other program on it that would erase it a line at a time.' The work 'was almost immediately pirated,' however, and exists as 'this permanent ghostly presence on the inter-net' (Gibson 8).
11 In his diary in *A Portrait*, Stephen writes of his 'sudden gesture of a revolutionary nature. I must have looked like a fellow throwing a handful of peas into the air' (*P* 252).
12 That 'blame' appears twelve times in the novel is an important fact: this number, that of the adventures of Homer's hero, crops up at unexpected moments in *Ulysses*. Groden (*Ulysses in Progress* 206) observes in the Rosenbach Manuscript a marginal note for Joyce's typist on the familiar

calling out in 'Telemachus.' For 'Steeeeeeeeeeephen' (*U-V* 20), 'N.B. There are <u>12 e's</u> here' (folio 26). However, editors and translators play variations: there are only ten in 'Steeeeeeeeephen' (*U* 24 and *U-M* 21) and eight in 'Il mio spirito familiare, dietro di me, che chiama Steeeeeeeephen' (de Angelis 21; this despite the translation's reliance upon Gabler, who gives twelve *es*). The Rosenbach Manuscript also punctuates this call with an exclamation mark, which only Rose's edition does (*U-RE* 21).

13 Women's rushed and rushing writing in *Ulysses* is a well-known and frequently discussed phenomenon, but Milly's 'infantile epistle' is sometimes overlooked, though it is interesting that it has a 'signature with flourishes' but 'no stop' (*U* 849).

14 The subjectivity of reading Joyce is discussed in chapter 7.

15 A 1916 reader's report of the manuscript of *A Portrait* complained that 'at the end of the book there is a complete falling to bits; the pieces of writing and the thoughts are all in pieces and they fall like damp, ineffective rockets ... this MS. wants time and trouble spent on it, to make it a more finished piece of work' (*L II* 371–2). Pound found the last phrase asinine, and after suggesting sending 'publishers readers to the serbian front, and get some good out of the war,' warned with strange prescience: 'Serious writers will certainly give up the use of english altogether unless you can improve the process of publication' (*L II* 373).

16 Michael Groden distinguishes between Joyce's having 'finished' *Ulysses* and having 'to stop writing' it (*Ulysses in Progress* 13, 200), and A. Walton Litz suggests that the novel 'provides a perfect illustration of Paul Valéry's remark that a work of art is never finished, but only abandoned' (*Art of James Joyce* 7).

17 'For Stephen art was neither a copy nor an imitation of nature: the artistic process was a natural process ... To talk about the perfection of one's art was not for him to talk about something agreed upon as sublime but in reality no more than a sublime convention but rather to talk a veritably sublime process of one's nature which had a right to examination and open discussion' (*SH* 154).

18 Rose hyphenates 'Monkey-doodle' (*U-RE* 115). The strange word may be an associational reflex on Bloom's part with Monks, the aging pressman who figures as an early Bloom doppelgänger. In 'Aeolus' Donald Theall observes a 'reduction of people and events to "types"' (77).

19 'Bull's eye!' cries Lynch to Stephen's first Aquinan definition; and then 'Bull's eye again!' to the next, 'wittily' (*P* 212), but he has no similar comeback for *claritas*, the most intangible quality of beauty, and one that, I argue in this chapter, Joyce's later work celebrates above wholeness (*integritas*) and harmony (*consonantia*).

20 I will have more to say about the function and effect of riddles and questions in Joyce's writing in chapter 9.
21 In the Linati Schema, the symbols of 'Aeolus' are of forms of desperation ('Hunger' and 'Failed Destinies'), but the Press connotes 'Mutability' (qtd in Groden, Ulysses *in Progress* 100).
22 It might be argued that the first words of *Ulysses*, in its first edition, are not 'Stately, plump,' but those of Sylvia Beach: '*The publisher asks the reader's indulgence for typographical errors unavoidable in the exceptional circumstances.*'
23 Even Fritz Senn has employed 'waiting for our redeemer' rhetoric in reference to *Ulysses* ('Prodding Nodding Joyce' 583).

4: Multiple Joyce Questions

1 A more mundane example of this mythologizing can be found in Suzette A. Henke's criticism of Eugene Jolas's 'presenting Joyce as an Irish surrealist' in his contribution to *Our Exagmination*. 'Words, for Joyce,' writes Henke, 'were never free-floating and "at liberty"; writing was not an automatic system welling up from the unconscious' (70). This is the determinist's Joyce: autocratic and even 'Godlike in his role as fabulator and linguistic fabricator' (Henke 70). Readers of *A Portrait* have heard it all before.
2 Even after the 'scandal of *Ulysses*,' the vilifications of Joyce continue. John Ralston Saul, for example, accusingly claims that Joyce, 'in his messianic fervour, was wholly aware that he was wrenching away the novel from the public – for whom it had been invented – and delivering it to the literary experts' (*Voltaire's Bastards: The Dictatorship of Reason in the West* [Toronto: Penguin, 1993], 560). Joyce himself appears on the cover of Saul's book, along with other notable bastards, dressed as a Western gunslinger.
3 My knowledge of Bishop's intriguing research comes from attending a talk entitled 'From Collectors to Cowpunchers: Who Bought *Ulysses*?' given at the 1998 Rome Joyce symposium.
4 The perspective of Joyce as author that I offer here is admittedly as distorted as any other, but I hope my critical lenses are appreciated more for their 'collideorscape' (*FW* 143.28) defamiliarizations than reverence.
5 I direct the reader to Maurice Blanchot's discussion of art and suicide in *The Space of Literature* (*L'espace littéraire*) (87–107). In Blanchot's analysis, the writer who writes rather than suicides is not necessarily making a consensual affirmation of his or her own agency or authority: 'Cet optimisme inconséquent qui rayonne à travers la mort volontaire, cette assurance de pouvoir toujours triompher, à la fin, en disposant souverainement du néant, en étant créateur de son propre néant, et, au sein de la chute, de pouvoir se hisser encore à la cime de soi-même, cette certitude affirme

dans le suicide ce que le suicide prétend nier' (128). (This illogical optimism which shines through voluntary death – this confidence that one will always be able to triumph in the end by disposing sovereignly of nothingness, by being the creator of one's own nothingness and by remaining able, in the very midst of the fall, to lift oneself to one's full height – this certitude affirms in the act of suicide the very thing suicide claims to deny [Smock 102–3].) The alternative that Joyce rejects is, in Bloom's words (thinking of his father's suicide), 'No more pain. Wake no more. Nobody owns' (U 121). The last sentences are wistful negations of both the title of Joyce's last book and his 'ownership' of it, negations that both Joyce and Blanchot understand are impossible.

6 Stillinger goes on to contend, however, that this myth is 'an absolute necessity' for criticism and interpretation, illogically drawing on Barthes for support (187). The absence of a postulated author is not the same as the postulation of a decentralized or multiple author.

7 Margot Norris observes a similar treatment of Ibsen in *Stephen Hero*: '[u]nread, unstaged, and therefore robbed of social impact, Ibsen becomes to young Dublin intellectuals an item of trivial pursuit in their parlor game of "Who's Who?"' (*Joyce's Web* 49).

8 In his 1964 novel, *The Dalkey Archive*, Flann O'Brien has his opportunistic protagonist, Mick, discover that James Joyce is not dead but tending a bar by the Irish seaside, living incognito and apparently without memory of his having written his last two books. Asked about *Ulysses*, he replies:

> –I paid very little attention to it until one day I was given a piece from it about some woman in bed thinking the dirtiest thoughts that ever came into the human head. Pornography and filth and literary vomit, enough to make even a blackguard of a Dublin cabman blush. I blessed myself and put the thing in the fire.
> –Well was the complete *Ulysses*, do you think, ever published?
> –I certainly hope not. (O'Brien 167)

9 Joyce's exact phrasing here, concerning 'who' is writing the 'crazy book,' is itself an interesting source of dispute. Hugh Kenner cites it as 'It is you, and you, and you, and that man over there, and that girl at the next table' (*Dublin's Joyce* 327), while Eric McLuhan, purportedly quoting Kenner, offers a condensed version: 'It is you, and you, and that girl at the next table' (17).

10 See André Topia, 'The Matrix and the Echo: Intertextuality in *Ulysses*' (Attridge and Ferrer 103–25).

11 Joyce's supervision of his official biographer is of interest here, particularly since Herbert Gorman's book has so many mistakes about its subject.

Bernard Benstock describes this biography as 'a melange of supportable facts, half-truths, and fictions that Joyce thought politic to plant at the time, and errors of fact or interpretation committed by Gorman that Joyce did not consider worth correcting' (75).

12 One resilient myth is of the Joyce who knew Greek. Campbell and Robinson list Greek among 'Sanskrit, Gaelic, and Russian' as the languages with which Joyce 'was on terms of scholarly intimacy' (358). This tendency to enhance the author's attributes is common – one finds it even in Borges (*Seven Nights* 119) – but Ellmann states unequivocally that Joyce 'knew no Greek' (118).

13 In a 1993 article, John Harty vehemently rejects the idea that 'Come in' remained in the *Wake*. Without consulting drafts, notebooks, or published instalments of 'Work in Progress,' Harty finds four instances of 'come in' and, surveying a (paraphrased) context for the respective passage in which each appears, concludes that they are syntactically 'integral' and 'not the result of a random accident' (54). The reasoning here favours the thesis that Joyce would rather cut or omit an anomaly than adapt the textual ground upon which it falls (or even the anomaly itself: try 'kommen' [*FW* 437.30]) to incorporate it. Since the composition of the *Wake* is very largely assimilative, Harty's project only casts doubt on Beckett's anecdote. The matter of Joyce's 'singular method' is corroborated by other accounts, including that of Eugene Jolas. Speaking of the arduous 'Work in Progress' proofing sessions, Jolas recounts how 'Joyce would improvise whenever something particularly interesting occurred to him during the reading, and occasionally even allow a *coquille* – a typographical error – to stand, if it seemed to satisfy his encyclopedic mind, or appeal to his sense of grotesque hazard' (Jolas 99).

14 The mystery surrounding 'A Little Cloud' remains, but readers will find my summary of and proffered solution possibility to it in *Notes and Queries* ('Sophoclean Cloudbusting in *Dubliners*'). Joyce's original title for 'Clay' was 'The Clay' (see *L II* 109, 111).

15 Mahaffey points out that the word '"text" derives from *texere*, to weave' (*Reauthorizing Joyce* 146).

16 'Casting such shadows to Persia's blind! The man in the street can see the coming event' (*FW* 583.14–15).

5: Fickling Intentions (I)

1 Joyce's relations with editors, so tempestuous in his *Dubliners* experience, eventually developed into a source of bemusement. While assembling *Finnegans Wake*, he wrote to Harriet Shaw Weaver: 'I am having queer

experiences with editors. New press opinions of Δ [Anna Livia Plurabelle] are: "all Greek to us" "unfortunately I can't read it" "is it a puzzle?" "has anybody had the courage to ask J. how many misprints are in it" "those French printers!" "how is your eyesight?" "charming!" – This last from Mrs. Nutting, who, however, heard me read it and indeed suggested my voice should be dished (misprint for "disced" [recorded])' (*L III* 131).

2 Cf. *A Portrait*: '[Stephen] smiled as he thought of the god's image [that of Thoth, the god of writers] for it made him think of a bottlenosed judge in a wig, putting commas into a document which he held at arm's length and he knew that he would not have remembered the god's name but that it was like an Irish oath. It was folly' (*P* 225).

3 Obviously, my use of the word 'text' in this book does not conform to its general usage within semiotics (though it may be closer to Barthes's usage than to that of Kristeva).

4 By way of digression: I have always been suspicious of the hyphen in the word 'new-comer' (*D* 48, *D-C* 49), which, though present in every edition of the book I have come across, does seem unlikely from the coiner of 'snotgreen' et al.

5 Andrew Goodwyn also makes this strange claim about *Dubliners*: 'the fact that the stories are interwoven, rather like a tapestry, makes it most unusual' (*D-C* 5). It is hard to say what Goodwyn means by 'interwoven' – the characters and plots stay confined within their own stories, after all – but another critically renowned fiction collection, Sherwood Anderson's *Winesburg, Ohio*, published within a few years of *Dubliners*, with which it is rarely compared, is no more or less 'unusual' and shares the collective bond (or weave) of localized place and time.

6 It seems incongruous that so abstract an image as Brancusi's 'Symbol of Joyce' (1929) is reproduced for the cover of Rose's *Reader's Edition*, but then again, John Joyce's wry response to the picture – 'The boy seems to have changed a good deal' (qtd in Ellmann 614) – holds a sentiment echoed by many of the edition's disenchanted reviewers. At least in the North American edition, Senn's and Groden's blurbs are absent from the paperback.

7 This statement, with which I concur, does seem to contradict another of Gottfried's pronouncements, cited in chapter 3, about Joyce's not being 'really interested in the process of printing' (64).

8 In a review entitled 'How Molly Bloom Got Her Apostrophes,' Lawrence Rainey offers the most vitriolic response to such changes: 'Rose likes punctuation because it appeals to his deep desire for order and facilitates what he calls "the undisturbed flow of the text" ... a desire that is thwarted by Molly Bloom's soliloquy' (594). Rainey concludes that Rose's 'edition, if

it can be called that, is a chastening example of how an excess of piety can imperceptibly turn into self-aggrandizing fantasy' (596).
9   The French translation that Joyce himself oversaw does not approximate this 'error': 'livre de style médiocrement littéraire intitulé *Les Douceurs du Péché*' (Morel 981).
10 An important question here concerns what form the project will take: CD-ROM, Web-based (whatever may come). Because Web-based data as a rule are not independent of other data – cross-listings and links, many of which are literally the coincidental results of search engines, make this so – designers of a Web-based *Ulysses* would have to decide how much interaction with other sites would be to their advantage. Annotations to Joyce's use of *Hamlet*, for instance, or references to Parnell, could offer, directly or as a secondary option, links to sites specifically focused on these subjects. This arrangement would prevent the need for the time and energy spent in constructing what would have to be an inferior, because much more limited, database on these tangential subjects, but it requires a firm trust in the quality and stability of the linked sites.
11 This is not at all an original definition on my part, but a tweaking of Hans Georg Gadamer's advancement of the idea 'that a text magnetizes on it, so to speak, the whole of the readings it has elicited in the course of history' (Eco, *Limits of Interpretation* 12). Further discussion of my highlighting 'mistakes' as the nexus of this history-in-progress occurs in Part III.

6: (Sic) of irony

1 Here, I am naturally thinking of Alanis Morrissette's 1995 song, 'Ironic.' To the dismay of teachers of literature everywhere, none of the scenarios described in the lyrics (e.g., 'It's like rain on your wedding day') is ironic; thus, only the title of the song is so. As for television, I direct the reader to David Foster Wallace, 'E Unibus Pluram: Television and U.S. Fiction' (*A Supposedly Fun Thing I'll Never Do Again: Essays and Arguments* [Boston: Little, Brown, 1997] 21–82).
2 This resolute but failed effort by the New Critics to address directly, to outstare 'the pitiless (though perhaps a bit strabismic) gaze of irony' (Booth 196) may well have informed the method of neophyte proponents of Deconstruction. Irony became not a subject but a demonstrable style, which in turn became a fashion. (*Fashion*, it is worth noting in the context of this dis-eased chapter, shares the same etymological roots as *infection*.) Accordingly, 'errant' deviations of form and language became useful. Ernst Behler contemplates Derrida's 'différance' as 'a visible, yet inaudible

spelling error' but the '*a* in the title and following usage of the monstrous word is ... no printing error, but a deliberate infusion by Derrida to make difference differ more than it normally does' (114). Behler's puzzling use of 'normally' probably denotes 'difference' as a non-textual idea or experience.

3 'Parallax' is listed among the 'characters' for 'Lestrygonians' in Joyce's notes (Litz 24). It is hard to imagine how this early scheme may have played out in the book (less so than, say, Milton's original plan to have 'Conscience' as an active character in *Paradise Lost*, since that work employs an appreciable allegorical framework).

4 One finds an equally pertinent understanding of irony expressed in Vico's *The New Science*: 'Irony certainly could not have begun until the period of reflection, because it is fashioned of falsehood by dint of a reflection which wears the mask of truth. Here emerges a great principle of human institutions, confirming the origin of poetry ... the first fables could not feign anything false; they must therefore have been ... true narrations' (131). Whatever one thinks of Vico as a historian (and for him the job description is one who investigates myth), his appreciation of poetry as at least originally 'true' is understandable inspiration for *Finnegans Wake*, the apocalypse of linguistic and textual norms.

5 Joseph A. Dane opens his study of irony with a nod to the 'interesting misprint under the entry "Irony"' in Johnson's *Dictionary*, where the Greek source *eirōneia* appears as *ierōneia* (1).

6 No edition can seem to agree with any other as to how to set this passage. To those cited in the course of the chapter I can add the case of the Modern Library edition:

> Rich booty you brought back; *Le tutu*, five tattered numbers of *Pantalon Blanc et Culotte Rouge*, a blue French telegram, curiosity to show:
>
> –Mother dying come home father.
> The aunt thinks you killed your mother. That's why she won't. (*U-M* 43)

7 I am indebted to Ed Germain for these ideas about Morse code.

8 Such frustrations may be behind the *Wake* sentence, 'Morse nuisance noised' (*FW* 99.06).

9 The navigational fastidiousness of 'Eumaeus' demands that Molly's muddled word be marked wrong: 'the book about Ruby with met him pike hoses (*sic*) in it' (*U* 760).

10 The edition of *Ulysses* cited here has an interesting anomaly in Stephen's first pronunciation of this thesis:

> –Where there is a reconciliation, Stephen said, there must have been first a sundering. (*U* 247)

The extra space between 'reconciliation,' and 'Stephen' is retained from the original Bodley Head editions (to which set it appears to be exclusive). It is tempting to read this space as a signal of hesitation by Stephen, especially since his point here, so immediately and agreeably digested by his listeners, is (literally) overstated and his weakest. Be that as it may, the notion of reconciliation's contingency upon sundering serves both as a caveat to editors and as an encouragement. Danis Rose makes some minor but unusual 'reconciliations' to this scene:

> –If that were the birthmark of genius, he said, genius would be a drug in the market. The plays of Shakespeare's later years, which Renan admired so much, breathe another spirit.
> –The spirit of reconciliation, the Quaker librarian breathed.
> –There can be no reconciliation, Stephen said, if there has not been a sundering. Said that. (*U-RE* 186)

11 In his own way Alan Wilde's delineations of irony are just as normative and dependent upon stability as those of Booth. In *Horizons of Assent*, Wilde tries to drive a wedge between modernism and postmodernism, contrasting the former's 'disjunctive irony' that 'strives, however reluctantly, toward a condition of paradox' and 'both recognizes the [world's] disconnections and seeks to control them' (10) with the latter's 'suspensive' irony, wherein 'an indecision about the meanings or relations of things is matched by a willingness to live with uncertainty, to tolerate and, in some cases, to welcome a world seen as random and multiple, even, at times, absurd' (44). Joyce's interplay between the modernisms, we might say, creates 'suspensive disjunctions'; but then again, as Buck Mulligan observes, '[w]e have grown out of Wilde and paradoxes' (*U* 21).

12 In the conclusion to their *Skeleton Key*, Campbell and Robinson oppose the dismissal of Joyce as a 'decadent' artist: 'If Joyce is sick, his disease is the neurosis of our age' (363).

7: Performance Anxieties

1 Maybe more as a personal fancy than a supportable argument, I suggest that the maddening 'U.P.' postcard may be the news, '*Ulysses* published,' and Breen (who, like so many other characters in the novel, could be a portrait of a contemporary of Joyce) in his fury foretells all other cases of indignation at the dear dirty Dubliners found in the novel. Breen's struggle with his own textual representation is perhaps suggested by his name: the *OED* recognizes *breended* as an obsolete form of *brinded*, 'the sense [of which] appears to be "marked as by burning" or "branding."'

2 I don't wish to misrepresent Bishop, whose *Joyce's Book of the Dark* is not at all indicative of these trends, and the above passage is chosen for its anomalous nature within that volume.
3 Bishop, who wrote the introduction to this particular reissue, may have provided the outline.
4 Iser's use of the word 'text' is at times confusing, alternating as it sometimes does with 'work.' For example, his claim that '[t]he work is more than the text, for the text only takes on life when it is realized' (274) would suggest that the 'inexhaustibility' he refers to later ought to be attributed not to the text but the work.
5 For a more detailed version of Eco's critique of Hermeticism see *Interpretation and Overinterpretation* 29–34.
6 Note that the title is different on the recording cited in the bibliography.

8: Fickling Intentions (II)

1 Robert Graves also has problems with Joyce's spelling of this name (see appendix) and these kinds of difficulties prompted Pound to write to Joyce in 1917: 'Since you will get yourself reviewed in modern Greek and thereby suggest new spellings of the name Daedalus. All I can say is Echt Dzoice, or Echt Joice, or however else you like it' (*L II* 413).
2 Joyce's paronomasia and textual derangements are not affirmations of *langue* but, on the contrary, his trademark contribution to a poetics of struggle with a top-down word of law. This poetics can be seen evolving in the resistance to a grammar expressed in Nietzsche and Blake and in the suspicion of vocabulary's agency in the works of Jack Spicer and various sound poets.
3 Pages 120–1 of the *Wake* have many other walk-on cameos by members of the alphabet: 'Greek ees' (120.19), 'the fretful fidget eff' (120.33), 'and the geegees too' (120.20–1), 'disdotted aiches' (121.16), 'doubleyous' (120.28), and so forth.
4 The sequential editing error in James's *The Ambassadors*, which I mentioned in Part I, allows for two slightly divergent reading paths through the novel. What interesting, erroneous impressions of 16 June 1904 would one have, I wonder, from a backwards reading of *Ulysses*?
5 Norris makes two serious errors here. The first is a misquotation, since the *Wake* passage actually reads: 'thence must any whatyoulike [or 'what-youlike'] in the power of empthood be either greater THaN or less THaN the unitate' (*FW* 298.11–3: observe the 'a'). Second, it is not completely true that 'any number,' n, raised to the power zero equals one; only real n, where $n < 0 > n$.

6 Bishop notes Joyce's mention of 'astronomical telescopes' (*L I* 235) and writes that such telescopes, 'unlike regular ones, work only at night, and they train on matters invisible to the light of day; they do what Joyce does in "his book of the dark"' (Bishop 21). *Finnegans Wake* attributes this ability not only to itself: 'When I'm dreaming back like that I begins to see we're only all telescopes' (*FW* 295.10–12).

9: The allriddle of it

1 Some question mark calculation trivia, based upon my own tallying: in *Finnegans Wake* the page with the most marks is 89 (twenty-six marks), and the most concentrated cluster occurs in pages 88–90 (sixty-one in three pages). About two-thirds of the pages of the *Wake* (418 of them) bear question marks. Of course, the punctuation's exuberance is borne out in other 'points': for instance, the *Wake* has prompted John Updike to wonder: 'has any book ever had so many exclamation points?' (137).
2 Brancusi's portrait, 'Symbol of Joyce,' was executed three years before. While the spiral and vertical lines are, as Guy Davenport suggests, reminiscent of the labyrinth of an ear (*Geography* 53), there is, too, the aspect of an exploded question mark.
3 As often as not this second use of questioning appears in a language of privilege, further emphasizing the imbalance of power between interrogator and interrogated. Joyce's indignation in his 1907 essay, 'Ireland at the Bar,' concerning 'the Irish question' and 'erroneous judgment' (*CW* 199) in the case of Myles Joyce, is palpable: '[t]he questioning, conducted through the interpreter, was at times comic and at times tragic ... [t]he figure of this dumbfounded old man ... deaf and dumb before his judge, is a symbol of the Irish nation at the bar of public opinion' (*CW* 197–8).
4 'Clearobscure' (*FW* 247.34) Joyce would laugh up his sleeve, however, at the convention of Irish scribes' placing 'more emphasis on the visual impact of punctuation and layout, which produced greater clarity' (Parkes 25).
5 This 'mark' can also be 'the curious warning sign' (121.08). Joyce is in part referring to the structure of his notebooks, where sigla (e.g., Issy's $\perp$) and letters (in the British Library's Notebook 47471B, for example, one finds W, B, M) act as markers placing supplementary text.
6 When the director counsels Stephen, '[y]our catechism tells you that the sacrament of Holy Orders is one of those which can be received only once because it imprints on the soul an indelible spiritual mark which can never be effaced' (*P* 160), he employs a metaphor that connects vocation with permanence in the same style as one that connects sin or fallenness with

permanence (both the mark of Cain and the typographical error are relevant here).

7 Elsewhere the quiz-chaos grows distinctly quixotic: 'usking queasy quizzers of his ruful continence' (*FW* 198.35).

8 History, too, has these methodological roots. The Greek verb *historein* means 'to ask questions' and Herodotus accordingly characterized his writings as 'inquiries.'

9 Aristotle's authorship of this work has been disputed, but this 'questionable' quality lends *Problems* an even greater Joycean relevance: 'we must vaunt no idle dubiosity as to its genuine authorship and holusbolus authoritativeness' (*FW* 117.35–118.04).

10 Kenner finds the same impulse in as early and minor a work as 'The Holy Office': 'The poet pretends to be a public servant relieving the "timid arses" of the "mumming company" by actually performing the psychic drama towards which they gesture. But his real function is inquisitorial' (*Dublin's Joyce* 298).

11 Compare this choice of adjective with the dean's 'useful arts' in *A Portrait* (*P* 185).

12 Gabriel's bad luck asking questions of women is first signalled with his failure at being playful with Lily; his bad luck at answering the questions of women is apparent in his exchange with Molly Ivors: 'Perhaps he ought not to have answered her like that' (*D* 191).

13 'Le plaisir du texte est semblable à cet instant intenable, impossible, purement *romanesque*, que le libertin goûte au terme d'une machination hardie, faisant couper la corde qui le pend, au moment où il jouit' (Barthes, *Le plaisir du texte* 15: 'The pleasure of the text is similar to that intangible, impossible, purely *romantic* instant at which the libertine tastes at the end of a daring scheme, cutting the cord that suspends him, at the moment when he is pleasured' [my translation]). The inquisition of *Finnegans Wake*, like the gorgeous frustrations of Italo Calvino's *If on a winter's night a traveler*, seeks to keep the reader dangling.

14 In *Wittgenstein's Ladder*, Marjorie Perloff points out that Fish's notion of an 'interpretive community' is, for Wittgenstein, 'precisely what one sets oneself up against, artistic pleasure being of necessity a private experience prompted by an individual practice' (79).

15 In most of the instances where publishers and journalists call *Finnegans Wake* 'complex' or 'challenging,' these terms can be understood to mean simply 'not much read.' Certainly, it remains the province of the brave and/or foolhardy, but this aspect of the text informs, in fact shapes, its meaning (*Bedeutung*) as much as, say, the understanding that Homer's

Odyssey has a largely pretextual pedigree. Such qualities, I would argue, supersede interpretation as an individual act, because they are the life story of the text.

16 McHugh, in *Annotations*, cites H. Travers Smith's *Psychic Messages from Oscar Wilde*: 'I was always one of those for whom the visible world existed' (88).

17 In Richard Powers's 1995 novel, *Galatea 2.2*, a cognitive neurologist and a novelist develop an electronic neural network by teaching it to read the works on an English Literature Comprehensive Exam reading list. At one point the novelist considers what might be 'a book's only conceivable theme': 'Life remembered. Life described. It wrote down and repeated what worked. The small copying errors the text made in running off examples of itself, edited by the world's differential rejection and forgiveness, produced the entire collaborative canon' (214). These remarkable sentences trace the pulse of *Finnegans Wake*.

18 There are also reprehensible answers, probably the best example of which is Mr Deasy's response, 'Because she never let them in' (*U* 44). The throwaway question, 'Are you saved?' (*U* 190), offers a rather different angle to the persecution of Jews and resonates more forcefully in *Ulysses* because, unlike Deasy's question, it is unanswerable – at least, within the text, since the reader is the ultimate recipient of these questions.

Bibliography

Abrams, M.H., et al., eds. *The Norton Anthology of English Literature*. 7th ed. Vol. 2. New York: Norton, 2000.
Adams, Robert Martin. *Surface and Symbol: The Consistency of James Joyce's Ulysses*. New York: Oxford UP, 1962.
Anderson, Chester G., ed. *A Portrait of the Artist as a Young Man: Text, Criticism and Notes*. New York: Penguin, 1977.
Aristotle. *Poetics*. Trans. I. Bywater. *The Complete Works of Aristotle*. Ed. Jonathan Barnes. Vol. 2. Princeton: Princeton UP, 1984. 2316–40.
– *Problems*. Trans. E.S. Forster. *The Complete Works of Aristotle*. Ed. Jonathan Barnes. Vol. 2. Princeton: Princeton UP, 1984. 1319–527.
Atherton, James S. *The Books at the Wake: A Study of Literary Allusions in James Joyce's Finnegans Wake*. 1959. Mamaroneck, N.Y.: Appel, 1974.
Attridge, Derek. *Joyce Effects: On Language, Theory, and History*. Cambridge: Cambridge UP, 2000.
– *Peculiar Language: Literature as Difference from the Renaissance to James Joyce*. London: Methuen, 1988.
– ed. *The Cambridge Companion to James Joyce*. Cambridge: Cambridge UP, 1990.
Attridge, Derek, and Daniel Ferrer, eds. *Post-structuralist Joyce: Essays from the French*. Cambridge: Cambridge UP, 1984.
Bacon, Francis. *Novum Organum*. Trans. and ed. Peter Urbach and John Gibson. Chicago: Open Court, 1994.
Barthes, Roland. *Le plaisir du texte*. Paris: Éditions du Seuil, 1973.
– *Writing Degree Zero & Elements of Semiology*. Trans. Annette Lavers and Colin Smith. London: Jonathan Cape, 1967.
Bates, Robin. 'The Correction Officer: Can John Kidd Save *Ulysses*?' *Lingua franca* 7.8 (1997): 38–46.
Beach, Sylvia. *Shakespeare and Company*. New York: Harcourt Brace, 1959.

Beckett, Samuel. 'Dante ... Bruno. Vico .. Joyce.' Beckett et al. 1–22.
- *Proust* and *Three Dialogues with Georges Duthuit*. London: John Calder, 1999.
- *The Unnamable. Three Novels*. New York: Grove P, 1958. 289–414.
Beckett, Samuel, et al. *Our Exagmination Round His Factification for Incamination of* Work in Progress. 1929. London: Faber, 1961.
Begnal, Michael H., and Fritz Senn, eds. *A Conceptual Guide to* Finnegans Wake. Philadelphia: Pennsylvania State UP, 1974.
Behler, Ernst. *Irony and the Discourse of Modernity*. Seattle: U of Washington P, 1990.
Benjamin, Walter. 'On the Program of the Coming Philosophy.' Trans. Mark Ritter. *Selected Writings: Volume 1: 1913–1926*. Ed. Marcus Bullock and Michael W. Jennings. Cambridge: Harvard UP, 1996. 100–10.
Bennett, Deborah J. *Randomness*. Cambridge: Harvard UP, 1998.
Benstock, Bernard. *Joyce-Again's Wake: An Analysis of* Finnegans Wake. Seattle: U of Washington P, 1965.
Berman, Art. *Preface to Modernism*. Urbana: U of Illinois P, 1994.
Bishop, John. *Joyce's Book of the Dark:* Finnegans Wake. Madison: U of Wisconsin P, 1986.
Blanchot, Maurice. *L'espace littéraire*. Paris: Gallimard, 1955.
Booker, M. Keith. *Joyce, Bakhtin, and the Literary Tradition: Toward a Comparative Cultural Poetics*. Ann Arbor: U of Michigan P, 1995.
Booth, Wayne C. *A Rhetoric of Irony*. Chicago: U of Chicago P, 1974.
Borges, Jorge Luis. *Collected Fictions*. Trans. Andrew Hurley. New York: Viking, 1998.
- *Historia universal de la infamia*. Buenos Aires: Emecé, 1966.
- *Seven Nights*. Trans. Eliot Weinberger. New York: New Directions, 1984.
Bornstein, George, ed. *Representing Modernist Texts: Editing as Interpretation*. Ann Arbor: U of Michigan P, 1991.
Broch, Hermann. *James Joyce und die Gegenwart*. Frankfurt: Suhrkamp, 1972.
- *Der Tod des Vergil*. Frankfurt: Suhrkamp, 1976.
Browne, Thomas. *Pseudoxia Epidemica. The Major Works*. Ed. C.A. Patrides. London: Penguin, 1977. 163–259.
Buckalew, Ronald E. 'Night Lessons on Language.' Begnal and Senn 93–115.
Budgen, Frank. 'James Joyce's *Work in Progress* and Old Norse Poetry.' Beckett et al. 35–46.
Burgess, Anthony. *Here Comes Everybody: An Introduction to James Joyce for the Ordinary Reader*. London: Faber, 1965.
- ed. *A Shorter Finnegans Wake*. London: Faber, 1978.
Burke, Kenneth. *A Grammar of Motives*. Berkeley: U of California P, 1969.

Burrell, Harry. *Narrative Design in* Finnegans Wake: *The* Wake *Lock Picked.* Gainesville: UP of Florida, 1996.

Burroughs, William S., and Brion Gysin. *The Third Mind.* New York: Viking, 1978.

Butler, Christopher, ed. *The Ambassadors.* By Henry James. Oxford: Oxford UP, 1985.

Calasso, Roberto. *The Marriage of Cadmus and Harmony.* Trans. Tim Parks. New York: Vintage, 1993.

Calvino, Italo. 'Two Interviews on Science and Literature.' *The Uses of Literature: Essays.* Trans. Patrick Creagh. San Diego: Harcourt, 1986. 28–38.

Campbell, Joseph, and Henry Morton Robinson. *A Skeleton Key to* Finnegans Wake. New York: Viking, 1961.

Carroll, Lewis. *Alice's Adventures in Wonderland. The Complete Works of Lewis Carroll.* London: Penguin, 1988. 9–120.

Chomsky, Noam. *Syntactic Structures.* The Hague: Mouton, 1966.

Cixous, Hélène. *The Exile of James Joyce.* Trans. Sally A.J. Purcell. New York: David Lewis, 1972.

– 'Joyce: The (r)use of writing.' Trans. Judith Still. Attridge and Ferrer 15–29.

Colomb, George G., and Mark Turner. 'Computers, Literary Theory, and Theory of Meaning.' *The Future of Literary Theory.* Ed. Ralph Cohen. New York: Routledge, 1989. 386–410.

Conley, Tim. '"Oh me none onsens!" *Finnegans Wake* and the Negation of Meaning.' *James Joyce Quarterly* (forthcoming).

– 'Sophoclean Cloudbusting in *Dubliners.*' *Notes and Queries* 245 (2000): 339–40.

Dalton, Jack P. 'Advertisement for the Restoration.' Dalton and Hart 119–37.

Dalton, Jack P., and Clive Hart, eds. *Twelve and a Tilly: Essays on the Occasion of the 25th Anniversary of* Finnegans Wake. London: Faber, 1966.

Dane, Joseph A. *The Critical Mythology of Irony.* Athens: U of Georgia P, 1991.

Davenport, Guy. *Every Force Evolves a Form: Twenty Essays.* London: Secker and Warburg, 1987.

– *The Geography of the Imagination.* San Francisco: North Point P, 1981.

De Angelis, Giulio, trans. *Ulisse.* By James Joyce. Italy: Arnoldo Mondadori, 1991.

De Man, Paul. *Allegories of Reading: Figural Language in Rousseau, Nietzsche, Rilke, and Proust.* New Haven: Yale UP, 1979.

Derrida, Jacques. 'Two Words for Joyce.' Trans. Geoff Bennington. Attridge and Ferrer 145–58.

– 'Ulysses Gramophone: Hear Say Yes in Joyce.' Trans. Tina Kendall and Shari Benstock. *Acts of Literature.* Ed. Derek Attridge. New York: Routledge, 1992. 253–309.

Dettmar, Kevin J.H. 'The Joyce That Beckett Built.' *James Joyce Quarterly* 35 (1998): 605–19.
- '"Working in Accord with Obstacles": A Postmodern Perspective on Joyce's "Mythical Method."' Dettmar, ed. 277–96.
- ed. *Rereading the New: A Backward Glance at Modernism*. Ann Arbor: U of Michigan P, 1992.

Duck Soup. Dir. Leo McCarey. Perf. the Marx Brothers. Paramount, 1933.

Duncan, Robert. 'After a Long Illness.' *Poems for the Millennium*. Vol.2. Ed. Jerome Rothenberg and Pierre Joris. Berkeley: U of California P, 1998. 847–8.

Dunleavy, Janet Egleson, ed. *Re-Viewing Classics of Joyce Criticism*. Urbana: U of Illinois P, 1991.

Duszenko, Andrzej. 'The Relativity Theory in *Finnegans Wake*.' *James Joyce Quarterly* 32 (1994): 61–70.

Eco, Umberto. *Interpretation and Overinterpretation*. Ed. Stefan Collini. Cambridge: Cambridge UP, 1992.
- *Kant and the Platypus: Essays on Language and Cognition*. Trans. Alastair McEwen. New York: Harcourt, 2000.
- *The Limits of Interpretation*. Advances in Semiotics. Bloomington: Indiana UP, 1994.
- *The Open Work*. Trans. Anna Cancogni. Cambridge: Harvard UP, 1989.
- *Travels in Hyperreality*. Trans. William Weaver. San Diego: Harcourt, 1986.

Ellmann, Richard. *James Joyce*. 2nd ed. Oxford: Oxford UP, 1982.

Enright, D.J. *The Alluring Problem: An Essay on Irony*. Oxford: Oxford UP, 1986.

Ferrer, Daniel, and Michael Groden. 'Post-Genetic Joyce.' *The Romanic Review* 86 (1995): 501–12.

Fish, Stanley. *Is There a Text in This Class? The Authority of Interpretive Communities*. Cambridge: Harvard UP, 1980.

Fitch, Noel Riley. *Sylvia Beach and the Lost Generation: A History of Literary Paris in the Twenties & Thirties*. New York: Norton, 1983.

Ford, Hugh. *Published in Paris: A Literary Chronicle of Paris in the 1920s and 1930s*. New York: Macmillan, 1975.

Foucault, Michel. 'What Is an Author?' Trans. Josué V. Harari. *The Foucault Reader*. New York: Pantheon, 1984. 101–20.

Frechtman, Bernard, trans. *What Is Literature?* By Jean-Paul Sartre. London: Methuen, 1967.

Frehner, Ruth, and Ursula Zeller, eds. *A Collideorscape of Joyce: Festschrift for Fritz Senn*. Dublin: Lilliput, 1998.

Frost, Robert. 'On Extravagance: A Talk.' *Poetry and Prose*. New York: Holt, 1972. 447–59.

Froula, Christine. *To Write Paradise: Style and Error in Pound's* Cantos. New Haven: Yale UP, 1984.
Frye, Northrop. *Anatomy of Criticism: Four Essays*. 1957. Princeton: Princeton UP, 1990.
Gabler, Hans Walter. Afterword. *Ulysses: The Corrected Text*. By James Joyce. London: Bodley Head, 1986. 645–50.
– 'Danis Rose: Ulysses: A "Reader's Edition."' *James Joyce Quarterly* 34 (1997): 561–73.
– 'A Response to: John Kidd, "Errors of Execution in the 1984 *Ulysses*."' *Studies in the Novel* 22 (1990): 250–6.
– Rev. of *The Textual Diaries of James Joyce*, by Danis Rose. *James Joyce Quarterly* 33 (1996): 621–5.
– 'The Text as Process and the Problem of Intentionality.' *Text* 3 (1987): 107–16.
Gass, William H. *Finding a Form*. Ithaca: Cornell UP, 1996.
Gibson, William. 'An Eye on Tomorrow: Interview with William Gibson.' *Rampike* 11.1 (1999): 7–11.
Gifford, Don, with Robert J. Seidman. Ulysses *Annotated: Notes for James Joyce's* Ulysses. 2nd ed. Berkeley: U of California P, 1989.
Gilbert, Stuart. 'Prolegomena to *Work in Progress*.' Beckett et al. 47–76.
Gogarty, Oliver St. John. 'Roots in Resentment: James Joyce's Revenge.' *Observer*, 7 May 1939, 4.
Goodwyn, Andrew, ed. *Dubliners*. By James Joyce. New York: Cambridge UP, 1995.
Gottfried, Roy. *Joyce's Iritis and the Irritated Text: The Dis-lexic* Ulysses. Gainesville: UP of Florida, 1995.
Graves, Robert. *Lars Porsena, or The Future of Swearing and Improper Language*. London: Kegan Paul, Trench, Trubner; New York: Dutton, 1927.
Groden, Michael. 'Contemporary Textual and Literary Theory.' Bornstein 259–86.
– 'Perplex in the Pen – and in the Pixels: Reflections on *The James Joyce Archive*, Hans Walter Gabler's *Ulysses*, and "James Joyce's *Ulysses* in Hypermedia."' *Journal of Modern Literature* 22 (1998–9): 225–44.
– Ulysses *in Progress*. Princeton: Princeton UP, 1977.
Groden, Michael, et al., eds. *The James Joyce Archive*. 63 vols. New York: Garland, 1977–79.
Hart, Clive. 'Fritz in the Early Awning.' Frehner and Zeller 4–10.
Harty, John. 'Is Beckett's "Come in" in *Finnegans Wake*?' *Notes on Modern Irish Literature* 5 (1993): 52–6.
Hawthorne, Nathaniel. *The Scarlet Letter: A Romance*. New York: Penguin, 1986.
Hayman, David. *The 'Wake' in Transit*. Ithaca: Cornell UP, 1990.

- Ulysses: *The Mechanics of Meaning*. Rev. ed. Madison: U of Wisconsin P, 1982.
Heath, Stephen. 'Ambiviolences: Notes for Reading Joyce.' Attridge and Ferrer 31–68.
Heidegger, Martin. 'The Origin of the Work of Art.' *Deconstruction in Context: Literature and Philosophy*. Ed. Mark C. Taylor. Chicago: U of Chicago P, 1986.
Henke, Suzette A. 'Exagmining Beckett & Company.' *Re-Viewing Classics of Joyce Criticism*. Ed. Janet Egleson Dunleavy. Urbana: U of Illinois P, 1991. 60–81.
Herring, Phillip F. *Joyce's Uncertainty Principle*. Princeton: Princeton UP, 1987.
Hirsch, Jr, E.D. 'Objective Interpretation.' Newton-De Molina 26–54.
- 'Three Dimensions of Hermeneutics.' Newton-De Molina 194–209.
Hobbs, Jerry R. *Literature and Cognition*. Stanford: Center for the Study of Language and Information, 1990.
Hutcheon, Linda. *Irony's Edge: The Theory and Politics of Irony*. London: Routledge, 1994.
Iser, Wolfgang. *The Implied Reader: Patterns of Communication in Prose Fiction from Bunyan to Beckett*. Baltimore: Johns Hopkins UP, 1974.
Jackson, Marni. 'Dear Mr. Manning, You Can't Tell a Book by Its Cover Any More Than You Can Tell a Leader by His Looks.' *Globe and Mail* [Toronto], 4 Mar. 2000, D22.
James, Henry. 'Daisy Miller: A Study.' *The Turn of the Screw and Other Short Novels*. New York: Signet, 1962. 93–152.
Johnson, Tom. 'Failing: A Very Difficult Piece for String Bass.' Perf. Robert Black. *Bang on a Can Live*. Vol. 1. Composers Recordings, 1992.
Jolas, Eugene. *Man from Babel*. Ed. Andreas Kramer and Rainer Rumold. New Haven: Yale UP, 1998.
Joyce, James. *Chamber Music*. 1907. Levin, ed. 629–48.
- *The Critical Writings of James Joyce*. Ed. Ellsworth Mason and Richard Ellmann. New York: Viking, 1959.
- *Dubliners*. 1914. London: Penguin, 1992.
- *Finnegans Wake*. New York: Penguin, 1976, 1999.
- 'Gas from a Burner.' 1912. Levin, ed. 660–2.
- *Giacomo Joyce*. London: Faber, 1983
- 'The Holy Office.' 1904. Levin, ed. 657–60.
- *Letters of James Joyce*. Vol. 1. Ed. Stuart Gilbert. London: Faber, 1957.
- *Letters of James Joyce*. Vol. 2. Ed. Richard Ellmann. London: Faber, 1966.
- *Letters of James Joyce*. Vol. 3. Ed. Richard Ellmann. New York: Viking, 1966.
- *Pomes Penyeach*. 1927. Levin, ed. 649–56.
- *A Portrait of the Artist as a Young Man*. 1916. Anderson 5–253.
- *Stephen Hero*. Ed. Theodore Spencer, John J. Slocum, and Herbert Cahoon. St Albans, U.K.: Triad, 1977.

- *Ulysses*. London: Penguin, 1992.
- *Ulysses*. New York: Modern Library, 1942.
- *Ulysses: The Corrected Text*. Ed. Hans Walter Gabler with Wolfhard Steppe and Claus Melchior. London: Bodley Head, 1986.
- *Ulysses: A Facsimile of the First Edition Published in Paris in 1922*. Washington: Orchises, 1998.
- *Ulysses: A Reader's Edition*. Ed. Danis Rose. London: Picador, 1997.

Kafka, Franz. *Amerika*. Trans. Willa and Edwin Muir. New York: Schocken, 1996.

Kappel, Andrew J. 'Complete with Omissions: The Text of Marianne Moore's *Complete Poems*.' Bornstein 125–56.

Keats, John. 'On First Looking into Chapman's Homer.' *The Norton Anthology of Poetry*. 3rd ed., shorter. Ed. Alexander W. Allison et al. New York: Norton, 1970. 359.

Kenner, Hugh. *Dublin's Joyce*. London: Chatto, 1955.
- *Joyce's Voices*. Berkeley: U of California P, 1978.
- *The Pound Era*. Berkeley: U of California P, 1971.

Kierkegaard, Søren. *The Concept of Irony with Constant Reference to Socrates*. Trans. Lee M. Capel. London: Collins, 1966.

Knowlton, Eloise. *Joyce, Joyceans, and the Rhetoric of Citation*. Gainesville: UP of Florida, 1998.

Knox, Bernard. Introduction. *The Odyssey*. By Homer. Trans. Robert Fagles. New York: Penguin, 1996. 3–64.

Kundera, Milan. *Testaments Betrayed: An Essay in Nine Parts*. Trans. Linda Asher. New York: HarperCollins, 1993.

Lavergne, Philippe, trans. *Finnegans Wake*. By James Joyce. Paris: Gallimard, 1982.

Lecercle, Jean Jacques. *Philosophy of Nonsense: The Intuitions of Victorian Nonsense Literature*. London: Routledge, 1994.

Levin, Harry. *Memories of the Moderns*. London: Faber, 1980.
- ed. *The Portable James Joyce*. New York: Penguin, 1975.

Lewis, Wyndham. *Time and Western Man*. New York: Harcourt Brace, 1928.

Litz, A. Walton. *The Art of James Joyce: Method and Design in* Ulysses *and* Finnegans Wake. New York: Oxford UP, 1964.
- 'The Uses of the *Finnegans Wake* Manuscripts.' Dalton and Hart 99–106.

Llona, Victor. 'I Dont Know What to Call It But Its Mighty Unlike Prose.' Beckett et al. 93–102.

Madtes, Richard E. *The 'Ithaca' Chapter of Joyce's* Ulysses. Ann Arbor, Mich.: UMI Research P, 1983.

Mahaffey, Vicki. 'Intentional Error: The Paradox of Editing Joyce's *Ulysses*.' Bornstein 171–91.

- *Reauthorizing Joyce.* Cambridge: Cambridge UP, 1988.
McAlmon, Robert, with Kay Boyle. *Being Geniuses Together: 1920–1930.* New York: Doubleday, 1968.
McCarthy, Patrick. *The Riddles of* Finnegans Wake. Cranbury, N.J.: Associated UP, 1980.
- ed. *Critical Essays on James Joyce's* Finnegans Wake. New York: Hall; Toronto: Macmillan, 1992.
McGann, Jerome J. *Black Riders: The Visible Language of Modernism.* Princeton: Princeton UP, 1993.
- *A Critique of Modern Textual Criticism.* Charlottesville: UP of Virginia, 1992.
- *The Textual Condition.* Princeton: Princeton UP, 1991.
McGee, Patrick. 'The Error of Theory.' *Studies in the Novel* 22 (1990): 148–62.
- *Paperspace: Style as Ideology in Joyce's* Ulysses. Lincoln: U of Nebraska P, 1988.
McHugh, Roland. *Annotations to* Finnegans Wake. Rev. ed. Baltimore: Johns Hopkins UP, 1991.
- *The* Finnegans Wake *Experience.* Berkeley: U of California P, 1981.
- 'Recipis for the Price of the Coffin.' Begnal and Senn 18–32.
McLuhan, Eric. *The Role of Thunder in* Finnegans Wake. Toronto: U of Toronto P, 1997.
McLuhan, Marshall. *The Interior Landscape: The Literary Criticism of Marshall McLuhan 1943–1962.* ed. Eugene McNamara. New York: McGraw-Hill, 1969.
Melville, Herman. *Moby-Dick or, The Whale.* New York: Penguin, 1992.
Moncrieff, C.K. Scott, and Terence Kilmartin, trans. *The Remembrance of Things Past.* By Marcel Proust. 3 vols. New York: Random House, 1981.
Moore, Marianne. *Complete Poems.* New York: Penguin, 1991.
- 'Efforts of Affection.' *Complete Poems* 147.
- 'A Grave.' *Complete Poems* 49–50.
- 'The Mind Is an Enchanting Thing.' *Complete Poems* 134–5.
- 'Pedantic Literalist.' *Complete Poems* 37.
- 'Poetry.' *Complete Poems* 36.
- 'To a Snail.' *Complete Poems* 85.
- 'When I Buy Pictures.' *Complete Poems* 48.
Morel, Auguste, trans. *Ulysse.* By James Joyce. Paris: Gallimard, 1998.
Morris, Errol, dir. *A Brief History of Time.* Paramount, 1992.
Muecke, D.C. *The Compass of Irony.* London: Methuen, 1969.
Newton-De Molina, David, ed. *On Literary Intention.* Edinburgh: Edinburgh UP, 1976.
Norris, Margot. *The Decentered Universe of* Finnegans Wake: *A Structuralist Analysis.* Baltimore: Johns Hopkins UP, 1974.

– *Joyce's Web: The Social Unraveling of Modernism*. Austin: U of Texas P, 1992.
– 'The Postmodernization of *Finnegans Wake* Reconsidered.' Dettmar, ed. 343–62.
O'Brien, Flann. *The Dalkey Archive*. London: Flamingo, 1993.
Olson, Charles. *Call Me Ishmael*. 1947. *Collected Prose*. Ed. Donald Allen and Benjamin Friedlander. Berkeley: U of California P, 1997. 1–105.
Ortega y Gasset, José. 'The Misery and the Splendor of Translation.' Trans. Elizabeth Gamble Miller. *Theories of Translation: An Anthology of Essays from Dryden to Derrida*. Ed. Rainer Schulte and John Biguenet. Chicago: U of Chicago P, 1992.
Parkes, M.B. *Pause and Effect: An Introduction to the History of Punctuation in the West*. Aldershot, U.K.: Scolar P, 1992.
Paz, Octavio. *The Bow and the Lyre*. Trans. Ruth L.C. Simms. Austin: U of Texas P, 1965.
Penrose, Roger. *The Emperor's New Mind: Concerning Computers, Minds and The Laws of Physics*. London: Vintage, 1990.
Pepper, Thomas Adam. *Singularities: Extremes of Theory in the Twentieth Century*. Cambridge: Cambridge UP, 1997.
Perloff, Marjorie. *Wittgenstein's Ladder: Poetic Language and the Strangeness of the Ordinary*. Chicago: U of Chicago P, 1996.
Pinker, Steven. *The Language Instinct*. New York: HarperCollins, 1994.
Pontiero, Giovanni, trans. *The History of the Siege of Lisbon*. By José Saramago. San Diego: Harcourt, 1996.
Pound, Ezra. *A B C of Reading*. London: Faber, 1951.
– *The Cantos of Ezra Pound*. New York: New Directions, 1993.
Power, Arthur. *Conversations with James Joyce*. Ed. Clive Hart. London: Millington, 1974.
Powers, Richard. *Galatea 2.2*. New York: HarperCollins, 1995.
Proust, Marcel. *À la recherche du temps perdu*. Vols 2, 3. Paris: Gallimard, 1954.
Rabaté, Jean-Michel. 'Lapsus ex machina.' Trans. Elizabeth Guild. Attridge and Ferrer 79–101.
– 'Pound, Joyce and Eco: Modernism and the "Ideal Genetic Reader."' *Romanic Review* 86 (1995): 485–500.
Rainey, Lawrence. 'The Cultural Economy of Modernism.' *The Cambridge Companion to Modernism*. Cambridge: Cambridge UP, 1999. 33–69.
– 'How Molly Bloom Got Her Apostrophes.' Rev. of *Ulysses: A Reader's Edition*. Ed. Danis Rose. *James Joyce Quarterly* 34 (1997): 588–96.
Rasula, Jed. '*Finnegans Wake* and the Character of the Letter.' *James Joyce Quarterly* 34 (1997): 517–30.
– 'Indigence in the Archive.' *Postmodern Culture* 9.3 (1999).

Rasula, Jed, and Steve McCaffery, eds. *Imagining Language: An Anthology.* Cambridge: MIT P, 1998.

Reason, James. *Human Error.* Cambridge: Cambridge UP, 1990.

Rice, Thomas Jackson. *Joyce, Chaos, and Complexity.* Urbana: U of Illinois P, 1997.

Rogers, William Elford. *Interpreting Interpretation: Textual Hermeneutics as an Ascetic Discipline.* University Park: Pennsylvania State UP, 1994.

Rorty, Richard. 'The Pragmatist's Progress.' Eco, *Interpretation and Overinterpretation* 89–108.

Rose, Danis. Introduction. *Ulysses: A Reader's Editon.* By James Joyce. London: Picador, 1997. ix–xxxiii.

Sage, Robert. 'Before *Ulysses* – and After.' Beckett et al. 147–70.

Saramago, José. *História do Cerco de Lisboa.* Lisbon: Caminho, 1989.

Sartre, Jean-Paul. *Que'est-ce que la littérature?* Paris: Gallimard, 1967.

Schulze, Robin G. 'Textual Darwinism: Marianne Moore, the Text of Evolution, and the Evolving Text.' *Text* 11 (1998): 270–305.

Senn, Fritz. *Inductive Scrutinies: Focus on Joyce.* Dublin: Lilliput, 1995.

– *Joyce's Dislocutions: Essays on Reading as Translation.* Ed. John Paul Riquelme. Baltimore: Johns Hopkins UP, 1984.

– 'Prodding Nodding Joyce: The "Reader's Edition" of *Ulysses,* edited by Danis Rose: Some First Impressions.' *James Joyce Quarterly* 34 (1997): 573–83.

– 'Righting *Ulysses.*' *James Joyce: New Perspectives.* Ed. Colin MacCabe. Sussex: Harvester; Bloomington: Indiana UP, 1982. 3–28.

Shakespeare, William. *The Tragedy of Hamlet, Prince of Denmark. William Shakespeare: The Tragedies; The Poems.* London: Cambridge UP, 1986. 218–53.

– *Much Ado About Nothing. William Shakespeare: The Comedies; The Histories.* London: Cambridge UP, 1986. 124–47.

Silliman, Ron. *The Age of Huts.* New York: Roof, 1986.

Slingsby, G.V.L. 'Writes a Common Reader.' Beckett et al. 189–91.

Slote, Sam. 'Nulled Nought: The Desistance of Ulyssean Narrative in *Finnegans Wake.*' *James Joyce Quarterly* 34 (1997): 531–42.

Smock, Ann, trans. *The Space of Literature.* By Maurice Blanchot. Lincoln: U of Nebraska P, 1989.

Sokal, Alan, and Jean Bricmont, *Fashionable Nonsense: Postmodern Intellectuals' Abuse of Science.* New York: Picador, 1998.

Spenser, Edmund. *The Faerie Queene.* Ed. A.C. Hamilton. London: Longman, 1977.

Staples, Hugh B. 'Growing Up Absurd in Dublin.' Begnal and Senn 173–200.

Steiner, George. *After Babel: Aspects of Language and Translation.* 3rd ed. Oxford: Oxford UP, 1998.

- *Errata: An Examined Life*. London: Weidenfeld, 1997.
- *In Bluebeard's Castle: Some Notes towards the Redefinition of Culture*. New Haven: Yale UP, 1971.
- *No Passion Spent: Essays, 1978–1995*. New Haven: Yale UP, 1996.
- 'Work in Progress.' Rev. of *The Arcades Project*, by Walter Benjamin. *Times Literary Supplement* No. 5044, 3 December 1999, 3–4.

Stillinger, Jack. *Multiple Authorship and the Myth of Solitary Genius*. New York: Oxford UP, 1991.

Szondi, Peter. *On Textual Understanding and Other Essays*. Trans. Harvey Mendelsohn. Minneapolis: U of Minnesota P, 1986.

Tanselle, G. Thomas. *A Rationale of Textual Criticism*. Philadelphia: U of Pennsylvania P, 1989.

Taylor, Mark C. *Erring: A Postmodern A/theology*. Chicago: U of Chicago P, 1984.

Theall, Donald F. *Beyond the Word: Reconstructing Sense in the Joyce Era of Technology, Culture, and Communication*. Toronto: U of Toronto P, 1995.

Tindall, William York. *A Reader's Guide to* Finnegans Wake. London: Thames and Hudson, 1969.

Untermeyer, Jean Starr, trans. *The Death of Virgil*. By Hermann Broch. New York: Pantheon, 1945.

Updike, John. 'Simple-Minded Jim.' *Hugging the Shore: Essays and Criticism*. New York: Vintage, 1983. 129–39.

Vico, Giovanni Battista. *The New Science of Giambattista Vico*. 1744. Trans. Thomas Goddard Bergin and Max Harold Fisch. Ithaca: Cornell UP, 1968.

Vogler, Thomas A. 'Wonder did he wrote it himself: Meditations on Editing *Finnegans Wake* in the "Gabler Era."' *Studies in the Novel* 22 (1990): 192–215.

Weir, Lorraine. *Writing Joyce: A Semiotics of the Joyce System*. Bloomington: Indiana UP, 1989.

Wilde, Alan. *Horizons of Assent: Modernism, Postmodernism, and the Ironic Imagination*. Baltimore: Johns Hopkins UP, 1981.

Williams, William Carlos. 'A Point for American Criticism.' Beckett et al. 171–85.

Wimsatt, W.K., and M.C. Beardsley. 'The Intentional Fallacy.' Newton-De Molina 1–13.

Wittgenstein, Ludwig. *Philosophische Untersuchungen / Philosophical Investigations*. 3rd ed. New York: Macmillan, 1958.

Yeats, W.B. 'Long-Legged Fly.' *Collected Poems*. London: Picador, 1990. 381–2.

Index

Abin, César, 134
Adams, Robert, 47, 67–8, 128, 131, 159n4
Adorno, Theodor, 24, 147
Allen, Fred, 153
Anderson, Sherwood, 166n5
Aquinas, Thomas, 102, 158n6
Aristotle, 10, 138–9, 141, 172n9
Attridge, Derek, 109–10, 126–7

Bacon, Francis, 8–10, 157n3
Bakhtin, Mikhail, 56–7
Barthes, Roland, 10, 26–7, 121, 145, 164n6, 166n3, 172n13
Baudelaire, Charles, 55
Beach, Sylvia, 71, 127, 155
Beardsley, M.C., 61
Beckett, Samuel, 23, 26–7, 45–6, 52, 57, 94, 107, 130, 143, 149, 165n13
Beethoven, Ludwig van, 136
Behler, Ernst, 94, 167n2
Bell, Clive, 41
Benjamin, Walter, 24, 60, 83, 150
Benstock, Bernard, 48, 103–4, 109–11, 116, 164n11
Berkeley, George, 144–5
Berman, Art, 23–5, 33, 35

Bishop, Edward, 40, 163n3
Bishop, John, 106–7, 113, 155, 170nn2, 3, 171n6
Blake, William, 41, 44, 57–8, 77, 170n2
Blanchot, Maurice, 42, 132, 151, 163n5
Boccaccio, Giovanni, 136
Bohr, Niels, 9
Booker, M. Keith, 57
Booth, Wayne, 56, 81–3, 92, 169n11
Borges, Jorge Luis, 11, 44, 89, 101, 129, 159n4, 165n12
Boru, Brian, 89
Bosinelli, Rosa Maria, 155
Brancusi, Constantin, 166n6, 171n2
Breton, André, 62
Bricmont, Jean, 157n1, 160n3
Broch, Hermann, 23, 36, 82, 91
Brown, Terence, 65
Browne, Thomas, 26, 161n1
Bruno, Giordano, 24, 58, 107
Budgen, Frank, 45, 54, 102–3
Burgess, Anthony, 76, 128, 154
Burke, Kenneth, 92
Burroughs, William S., 63
Byron, Lord (George Gordon), 124

Index

Calasso, Roberto, 15
Calvino, Italo, 7, 172n13
Campbell, Joseph, 110–11, 113–14, 165n12, 169n12
Canning, A.S., 153
Carroll, Lewis, 44, 146
Cervantes, Miguel de, 34, 94; *Don Quixote*, 142, 159n4
Chomsky, Noam, 122
Cixous, Hélène, 138, 156
Colomb, George G., 142
Coltrane, John, 132
Confucius, 39
Cortázar, Julio, 95–6; *Hopscotch*, 124
Columbus, Christopher, 34

Dalton, Jack, 47
Dane, Joseph A., 92–3, 168n5
Dante Alighieri, 17, 34, 43, 89
Darwin, Charles, 28
Davenport, Guy, 39, 55, 115, 171n2
de Angelis, Giulio, 88
Defoe, Daniel, 67–8
Deleuze, Gilles, 159n10
de Man, Paul, 60, 92
Derrida, Jacques, 20, 57, 101, 105, 107, 135, 167n2
Descartes, René, 143–4
Dettmar, Kevin, 57
Dickinson, Emily, 27
Driver, Clive, 27
Dryden, John, 153
Duncan, Robert, 13
Duszenko, Andrzej, 160n3
Duthuit, Georges, 146

Eagleton, Terry, 19, 159n8
Eco, Umberto, 5, 83, 102, 105–6, 114–15, 124–7, 129, 131, 170n5
Edel, Leon, 50

Edgeworth, Francis Ysidro, 61
Eggers, Dave, 81
Einstein, Albert, 9
Eliot, T.S., 27, 41, 43, 48, 61, 65, 158n2
Ellmann, Richard, 45–6, 165n12
Empson, William, 82, 139

Feehan, Mike, 161n8
Fish, Stanley, 5, 123, 144, 172n14
Fitch, Noel Riley, 127
Flaubert, Gustave, 38
Foucault, Michel, 10
Frege, Gottlob, 11
Freud, Sigmund, 106–7
Frost, Robert, 34
Froula, Christine, 32–3, 39
Frye, Northrop, 82, 92

Gabler, Hans Walter, 16, 69–75, 85, 162n12
Gadamer, Hans Georg, 167n11
Galilei, Galileo, 9–10, 157n4
Genette, Gérard, 65
Germain, Ed, 168n7
Gibson, William, 161n10
Gifford, Don, 104
Gödel, Kurt, 24, 160n3
Gogarty, Oliver St-John, 19
Goodwyn, Andrew, 67, 166n5
Gorman, Herbert, 164n11
Gottfried, Roy, 19, 36, 71, 166n7
Graves, Robert, 156, 170n1
Groden, Michael, 19–20, 56, 69, 71, 75, 149, 161n12, 162n16, 166n6
Guattari, Félix, 159n10
Gysin, Brion, 63

Hart, Clive, 151, 155
Harty, John, 165n13
Hawking, Stephen, 9, 40, 157n5

Index

Hawthorne, Nathaniel, 26; *The Scarlet Letter*, 123–4
Hayman, David, 36, 51–6
Heath, Stephen, 43, 82
Heidegger, Martin, 10, 118
Heisenberg, Werner 9, 61
Henke, Suzette A., 163n1
Herodotus, 172n8
Herring, Phillip F., 48, 54
Hirsch, E.D., 60, 120–2
Hobbs, Jerry R., 62, 119–20
Homer, 15, 38, 110, 140, 161n12, 172n15
Howard, Philip, 81
Hubble, Edwin, 24
Hume, David, 143
Hutcheon, Linda, 92–3, 119

Ibsen, Henrik, 164n7
Iser, Wolfgang, 113, 122–3, 144, 170n4

James, Henry, 12, 23, 170n4; 'Daisy Miller,' 87; *The Golden Bowl*, 142
Johnson, George, 127, 129
Johnson, Samuel, 168n5
Johnson, Tom, 116–17
Jolas, Eugene, 43, 163n1, 165n13
Joyce, James: *Chamber Music*; 46, 53, 125; *Dubliners*, 36–7, 42, 47, 49–53, 63, 65–7, 88, 102, 105, 125, 141, 151, 153, 155–6, 165nn14, 1, 166nn4, 5, 172n12; *Exiles*, 141; 'Gas from a Burner,' 49; 'The Holy Office,' 49, 172n10; *Finnegans Wake*, 6, 8, 10, 15, 17–20, 25, 34–5, 37–9, 41–7, 49, 51, 53–6, 62–3, 67–8, 72–3, 75–80, 82–3, 85–6, 89, 91–4, 96–8, 101–17, 119–20, 122–5, 127–9, 131–47, 149, 151, 154–5, 159nn4, 8, 9, 160n3, 165nn13, 16, 165n1, 168n4, 170nn3, 5, 171n6, 172nn13, 15; *Pomes Penyeach*, 96, 125; *A Portrait of the Artist as a Young Man*, 11, 12, 14, 15, 35, 46, 51, 60, 83, 102, 105, 109–10, 124–5, 134, 137, 140, 149, 153–5, 158n6, 161n11, 162nn15, 19, 163n1, 166n2, 171n6, 172n11, 173n17; *Stephen Hero*, 60, 154, 162n17, 164n7; *Ulysses*, 6, 14–16, 20, 25, 33–45, 47–54, 57–8, 63–5, 67–77, 80, 82–3, 85–91, 94, 97–8, 101–2, 104–5, 109–10, 113, 115, 117, 120, 123, 126, 129–32, 134, 137, 141, 143–5, 151, 153–6, 158n2, 161nn6, 12, 162nn13, 16, 18, 163n2, 164n8, 166nn6, 8, 167n10, 168nn3, 6, 9, 10, 169n1, 170n4, 171n1, 172n7, 173n18
Joyce, John, 166n6
Joyce, Myles, 171n3
Joyce, Stanislaus, 46, 53, 89

Kafka, Franz, 12, 27
Kant, Immanuel, 143
Kappel, Andrew J., 27, 161n9
Kavanagh, Patrick, 98
Keats, John, 12, 153
Kenner, Hugh, 6, 30–1, 36, 51, 68, 143, 158n2, 164n9, 172n10
Kidd, John, 16, 69–70, 85, 89
Kierkegaard, Søren, 82, 85
Knowlton, Eloise, 17–18, 87–8, 154, 159nn4, 5
Kristeva, Julia, 166n3
Kundera, Milan, 34, 91–2

Lavergne, Philippe, 35
Leavis, F.R., 41
Levin, Harry, 112–13

Lewis, Wyndham, 41, 135
Litz, A. Walton, 108, 159n7, 162n16
Llona, Victor, 108
Locke, John, 143

Madtes, Richard E., 72–3, 139–40
Mahaffey, Vicki, 38, 61–3, 70, 72, 76, 165n15
Malebranche, Nicolas, 143
Mallarmé, Stéphane, 24
Mann, Thomas, 23, 93
Marinetti, F.T., 24
Maritain, Jacques, 144
Marx Brothers, 53, 137–8
McAlmon, Robert, 41, 46
McCaffery, Steve, 77
McCarthy, Patrick, 137–8, 145
McGann, Jerome, 33, 40, 42, 61, 63–4, 96
McGee, Patrick, 16, 52–4, 56, 70–1, 86, 158n2
McHugh, Roland, 17, 55–6, 80, 102–3, 106, 108, 110, 115, 121, 155, 159n4, 173n16
McLuhan, Eric, 110, 155, 164n9
McLuhan, Marshall, 6, 121, 150
Melville, Herman, 26–7, 29, 34–6, 161nn7, 8
Milton, John, 168n3
Moore, Marianne, 27–9, 34–6, 39, 54, 161n8
Morris, Errol, 157n5
Morrissette, Alanis, 167n1
Muecke, D.C., 82, 87, 91–3
Musil, Robert, 24, 91

Nabokov, Vladimir, 88
Nashe, Thomas, 153
New Yorker, 5
Nietzsche, Friedrich, 170n2

Norris, Margot, 19, 109, 112–13, 127–9, 137, 159n9, 164n7, 170n5
Norton Anthology of English Literature, The, 77, 79, 155

O'Brien, Flann, 164n8
Olson, Charles, 161n7
Ortega y Gasset, José, 157n2

Parkes, M.B., 135–6
Parnell, Charles Stewart, 167n10
Pascal, Blaise, 7, 59–60, 83
Pavlov, Ivan Petrovich, 145
Paz, Octavio, 32, 151
Peirce, Charles S., 114
Penrose, Roger, 6, 146
Pepper, Thomas, 60
Perec, Georges, 97, 143
Perloff, Marjorie, 172n14
Pessoa, Fernando, 47
Petrarch, 136
Pinker, Steven, 11
Pioneer 10 spacecraft, 62, 150
Planck, Max, 9
Plato, 37–8, 55, 92
Poe, Edgar Allan, 145
Pound, Ezra, 23, 29–36, 39, 158n7, 170n1
Power, Arthur, 35–6
Powers, Richard, 173n17
Proust, Marcel, 24, 27, 57, 63, 132–3, 136
Purcell, Henry, 131

Rabaté, Jean-Michel, 108, 114–15, 127, 143
Rabelais, François, 107
Rainey, Lawrence, 40–1, 166n8
Rasula, Jed, 19, 77, 143–4
Reason, James, 12

Rice, Thomas Jackson, 19–20
Richards, Grant, 47
Rimbaud, Arthur, 31, 39
Robbe-Grillet, Alain, 129
Robinson, Henry Morton, 110–11, 113–14, 165n12, 169n12
Rogers, William Elford, 121–2
Roosevelt, Franklin Delano, 31
Rorty, Richard, 104, 126
Rose, Danis, 16, 71–2, 74, 75, 86, 162n18, 166nn6, 8, 168n10
Roth, Samuel, 72
Russell, Bertrand, 24
Russell, George ('Æ'), 44

Sage, Robert, 48–9, 117
Salinger, J.D., 118
Saramago, José, 46–7
Sartre, Jean-Paul, 25, 160n4
Saul, John Ralston, 163n2
Saussure, Ferdinand de, 122
Schlegel, F.W., 60, 84, 85, 94
Schoenberg, Arnold, 160n1
Schulze, Robin G., 28–9
Senn, Fritz, 35, 69, 71, 74–6, 86, 94, 116, 154–5, 159n8, 166n6
Shakespeare, William, 44, 50, 61, 97, 154; *Hamlet*, 90–1, 131, 167n10; *Much Ado About Nothing*, 115–16; *Othello*, 91, 153
Silliman, Ron, 139, 143
Slingsby, G.V.L., 103, 131
Slote, Sam, 109–10
Smart, Christopher, 23
Smith, H. Travers, 173n16
Snow, C.P., 6
Socrates, 92, 115, 141, 145
Sokal, Alan, 157n1, 160n3
Spenser, Edmund, 157n3
Spicer, Jack, 170n2

Stein, Gertrude, 18, 27, 42, 124
Steiner, George, 7, 12–13, 24, 31–2, 63
Stephens, James, 45
Sterne, Lawrence, 23, 95; *Tristram Shandy*, 87, 124
Stillinger, Jack, 42, 164n6
Swift, Jonathan, 23, 65
Szondi, Peter, 147

Tanselle, G. Thomas, 64–5, 76, 80
Tatlin, Vladimir, 160n1
Taylor, Mark C., 159n10
Tennyson, Alfred, 50, 124
Thackeray, William, 41
Theall, Donald F., 110, 125, 162n18
Tindall, William York, 110–11, 112, 114
Topia, André, 164n10
Toynbee, Philip, 139
Turner, Mark, 142
Twain, Mark, 44
Tzara, Tristan, 24

Updike, John, 171n1

Valéry, Paul, 162n16
Verlaine, Paul, 43
Vico, Giambattista, 58, 107, 168n4
La vie en culotte, 89
Virgil, 61
Vogler, Thomas, 42, 154

Wagner, Richard, 118
Wallace, David Foster, 81, 167n1
Warburg, Aby, 63
Weaver, Harriet Shaw, 49, 51, 83, 108, 165n1
Weir, Lorraine, 143, 159n4
West, Rebecca, 108
Whitman, Walt, 27, 32, 161n8

Wilde, Alan, 82, 169n11
Wilde, Lady Jane Francesca
 ('Speranza'), 44
Wilde, Oscar, 44–5, 145, 169n11
Williams, William Carlos, 33, 108, 155
Wilson, Edmund, 139
Wimsatt, W.K., 61

Wittgenstein, Ludwig, 6, 13, 60, 62, 94, 112, 122, 139, 144, 172n14
Woolf, Virginia, 41–2

Yeats, William Butler, 113, 118–19, 153

Zukofsky, Louis, 33

www.ingramcontent.com/pod-product-compliance
Lightning Source LLC
Chambersburg PA
CBHW052027070526
44584CB00016B/1933